W9-BBT-692

How to Stay Smart in a Smart World

How to Stay Smart in a Smart World: Why Human Intelligence Still Beats Algorithms

Gerd Gigerenzer

The MIT Press
Cambridge, Massachusetts
London, England

The MIT Press would like to thank the anonymous peer reviewers who provided comments on drafts of this book. The generous work of academic experts is essential for establishing the authority and quality of our publications. We acknowledge with gratitude the contributions of these otherwise uncredited readers.

This book was set in ITC Stone Serif Std and ITC Stone Sans Std by New Best-set Typesetters Ltd. Printed and bound in the United States of America.

Library of Congress Cataloging-in-Publication Data

Names: Gigerenzer, Gerd, author.
Title: How to stay smart in a smart world : why human intelligence still beats
    algorithms / Gerd Gigerenzer.
Description: Cambridge, Massachusetts ; London, England : The MIT Press, [2022] |
    Includes bibliographical references and index.
Identifiers: LCCN 2021035121 | ISBN 9780262046954 (hardcover)
Subjects: LCSH: Artificial intelligence—Social aspects. | Expert systems (Computer
    science)—Safety measures. | Expert systems (Computer science)—Risk assessment.
Classification: LCC Q334.7 .G54 2022 | DDC 303.48/34—dc23
LC record available at https://lccn.loc.gov/2021035121

10 9 8 7 6 5 4 3 2 1

For Athena

Contents

Introduction

Daddy, when you were young, before computers, how did you get into the internet?
—A seven-year-old in Boston

If robots do everything, then what are we going to do?
—A five-year-old in a Beijing kindergarten, quoted in Lee, *Superpowers*

Imagine a digital assistant who does everything better than you. Whatever you say, it knows better. Whatever you decide, it will correct. When you come up with a plan for next year, it will have a superior one. At some point you may give up making any personal decisions on your own. Now it's the AI that efficiently runs your finances, writes your messages, chooses your romantic partner, and plans when it's best to have children. Packages will be delivered to your door containing products you didn't even know you needed. A social worker may turn up because the digital assistant predicted that your child is at risk of severe depression. And before you waste time agonizing over which political candidate you favor, your assistant will already know and cast your vote. It's just a matter of time until tech companies run your life, and the faithful assistant morphs into a supreme superintelligence. Like a flock of sheep, our grandchildren will cheer or tremble in awe of their new master.

In recent years, I have spoken at many popular artificial intelligence (AI) events and am repeatedly surprised at how widespread unconditional trust in complex algorithms appears to be. No matter what the topic was, representatives of tech companies assured listeners that a machine will do the job more accurately, more quickly, and more cheaply. What's more, replacing people by software might well make the world a better place. In the same vein, we hear that Google knows us better than we know ourselves and that AI can predict our behavior almost perfectly, or soon will be able to do so. Tech companies proclaim this ability when they offer their services to advertisers, insurers, or retailers. We too tend to believe it. Even those popular authors who paint doomsday pictures of robots ripping the guts out of humans assume the near omniscience of AI, as do some of the tech industry's most outspoken critics, who brand the business as evil *surveillance capitalism* and fear for our freedom and dignity.[1] It is this belief that makes many worry about Facebook (now renamed Meta) as a terrifying Orwellian surveillance machine. Data leaks and the Cambridge Analytica scandal have amplified this worry into fearful awe. Based on faith or fear, the story line remains the same. It goes like this:

AI has beaten the best humans in chess and Go.

Computing power doubles every couple of years.

Therefore, machines will soon do everything better than humans.

Let's call it the AI-beats-humans argument. It forecasts that a machine super-intelligence is near. Its two premises are correct, but the conclusion is wrong.

The reason is that computing power goes a long way for some kinds of problems but not for others. To date, the stunning victories of AI have been in well-defined games with fixed rules, such as chess and Go, with similar successes for face and voice recognition in relatively unchanging conditions. When the environment is stable, AI can surpass humans. If the future is like the past, large amounts of data are useful. However, if surprises happen, big data—which is always data from the past—may mislead us about the future. Big data algorithms missed the financial crisis of 2008 and predicted Hillary Clinton's victory by a large margin in 2016.

In fact, many problems we face are not well-defined games but situations in which uncertainty abounds, be it finding true love, predicting who will commit a crime, or reacting in unforeseen emergency situations. Here, more computing power and bigger data are of limited help. Humans are the key source of uncertainty. Imagine how much more difficult chess would be if the king could violate the rules at a whim and the queen could stomp off the board in protest after setting the rooks on fire. With people involved, trust in complex algorithms can lead to illusions of certainty that become a recipe for disaster.

To appreciate that complex algorithms are likely to succeed when situations are stable but will struggle with uncertainty exemplifies the general theme of this book:

Staying smart means understanding the potentials and risks of digital technologies, and the determination to stay in charge in a world populated by algorithms.

I wrote this book to help you understand the potential of digital technologies, such as AI, but even more important, the limitations and risks of these technologies. As these technologies become more widespread and dominant, I want to provide you with strategies and methods to stay in charge of your life rather than let yourself get steamrolled. Should we simply lean back and relax while software makes our personal decisions? Definitely not. Staying smart does not mean obliviously trusting technology, nor does it mean anxiously mistrusting it. Instead, it is about understanding what AI can do and what remains the fancy of marketing hype and techno-religious faiths. It is also about one's personal strength to control a device rather than being remote-controlled by it.

Staying smart is not the same as having digital skills for using technology. Educational programs worldwide seek to increase digital skills by buying tablets and smart whiteboards for classrooms and teaching children how to use them. But these programs rarely teach children how to understand the risks delivered by digital technology. As a consequence, most digital natives shockingly cannot tell hidden ads from real news and are taken in by the appearance of a website. For instance, a study of 3,446 digital natives showed that 96 percent of them do not know how to check the trustworthiness of sites and posts.[2]

A *smart world* is not just the addition of smart TVs, online dating, and gimmicks to our lives. It is a world *transformed* by digital technology. When the door to the smart world was first opened, many pictured a paradise where everyone had access to the tree of truthful information, which would finally put an end to ignorance, lies, and corruption. Facts about climate change, terrorism, tax evasion, exploitation of the poor, and violations of human dignity would be laid open. Immoral politicians and greedy executives would be exposed and forced to resign. Government spying on the public and violations of privacy would be prevented. To some degree, this dream has become reality, although the paradise has also been polluted.

What is really happening, however, is a transformation of society. The world does not simply get better or worse. How we think about good and bad is changing. For instance, not long ago, people were extremely concerned about privacy and took to the streets to protest against governments and corporations that tried to surveil them and get hold of their personal data. A wide spectrum of activists, young liberals, and mainstream organizations held massive protests against the 1987 German census, fearing that computers could deanonymize their answers, and angry people plastered the Berlin Wall with thousands of empty questionnaires. In the 2001 census in Australia, more than 70,000 people declared their religion as "Jedi" (after the movie *Star Wars*), and in 2011, British citizens protested against questions infringing on their privacy, such as about their religion.[3] Today, when our smart home records everything we do 24/7, including in our bedroom, and our child's smart doll records every secret it is entrusted with, we shrug. Feelings of privacy and dignity adapt to technology or may become concepts of the past. The dream of the internet was once freedom; for many, freedom now means free internet.

Since time immemorial, humans have created impressive new technologies that they have not always used wisely. To reap the many benefits of digital technology, we need the insight and courage to remain smart in a smart world. This is the time not to lean back and relax but to keep your eyes open and stay in charge.

Staying in Charge

If you are not a devil-may-care person, you might occasionally worry about your safety. Which disaster do you think is more likely to happen in the next ten years?

- You will be killed by a terrorist.
- You will be killed by a driver distracted by a smartphone.

If you opt for the terrorist attack, you are among the majority. Since the 9/11 attack, surveys in North America and Europe have indicated that many people believe terrorism poses one of the greatest hazards to their lives. For some, it is the greatest fear. At the same time, most admit to texting while driving without much concern. In the ten years before 2020, thirty-six people on average were killed annually in the United States by terrorists, Islamic, right wing, or other.[4] In that same period, more than 3,000 people were killed annually by distracted drivers—often people busy on their phone texting, reading, or streaming.[5] That figure amounts to the death toll of the 9/11 attack, but for every year.

Most Americans are also more afraid of terrorism than of guns, even though they are less likely to be shot by a terrorist than by a child playing with a gun in their household. Unless you live in Afghanistan or Nigeria, you will much more likely be killed by a distracted driver, possibly yourself. And it is not difficult to understand why. When twenty-year-old drivers use a phone, their reaction times decline to that of a seventy-year-old driver without one.[6] That's known as instant brain aging.

Why do people text while driving? They might not be aware of how dangerous it is. However, in a survey, I found that most are well aware that a hazard exists.[7] At issue is not lack of awareness. It is lack of self-control. "When a text comes in, I just have to look, no matter what," one student explains. And self-control has been made more difficult ever since platforms introduced notifications, likes, and other psychological tricks to keep users' eyes glued to their sites rather than their surroundings. Yet so much damage could be avoided if people managed to overcome their urge to check their phone when they should be paying attention to the road. And it's not just young people. "Don't text your loved ones when you know they are driving," said a devastated mom who found her badly injured daughter in the intensive care unit, face scarred and one eye lost, after having sent her child "a stupid text."[8] A smartphone is an amazing technology, but it requires smart people who use it wisely. Here, the ability to stay in charge and control a technology protects the personal safety of yourself and your loved ones.

Mass Surveillance Is a Problem, Not a Solution

Part of the reason why we fear a terrorist attack rather than a driver glued to a smartphone is that more media attention is devoted to terrorism than to distracted driving, and politicians have followed suit. To protect their citizens, governments all around the world experiment with face-recognition surveillance systems. These systems do an exceptional job of recognizing faces when tested in a lab using visa or job application photographs, or other well-lit photos with people's heads held in similar positions.

But how accurate are they in the real world? One test took place close to my home.

On the evening of December 19, 2016, a twenty-four-year-old Islamist terrorist hijacked a heavy truck and plowed into a busy Berlin Christmas market packed with tourists and locals enjoying sausages and mulled wine, killing twelve people and injuring forty-nine. The following year, the German Ministry of the Interior installed face-recognition systems at a Berlin train station to test how accurately they recognized suspects. At the end of the yearlong pilot, the Ministry proudly announced in its press release two exciting numbers: a hit rate of 80 percent, meaning that of every ten suspects, the systems identified eight correctly and missed two; and a false alarm rate of 0.1 percent, meaning that only one out of every 1,000 innocent passersby was mistaken for a suspect. The minister hailed the system an impressive success and concluded that nationwide surveillance is feasible and desirable.

After the press release, a heated debate arose. One group had faith that more security justifies more surveillance, while the other group feared that the cameras would eventually become the "telescreens" in George Orwell's *1984*. Both, however, took the accuracy of the system for granted.[9] Instead of taking sides in that emotional debate, let's consider what would actually happen if such face-recognition systems were widely implemented. Every day, about twelve million people pass through train stations in Germany. Apart from several hundred wanted suspects, these are normal people heading for work or out for pleasure. The impressive-sounding false positive rate of 0.1 percent translates into nearly 12,000 passersby per day who would be falsely mistaken as suspects. Each one would have to be stopped, searched for weapons or drugs, and be restrained or held in custody until their identity is proven.[10] Police-related resources, already strained, would be used for scrutinizing these innocent citizens rather than for effective crime prevention, meaning that such a system would in fact come at the cost of security. Ultimately, one would end up with a surveillance system that infringes on individual freedom and disrupts social and economic life.

Face recognition can perform a valuable service, but for a different task: *identification of an individual* rather than *mass screening*. After a crime happens at a subway station or a car runs through a red light, a video recording can help to identify the perpetrator. Here we know that the person has committed a crime. When screening everyone at the station, in contrast, we do not know whether the people being screened are suspects. Most of them aren't, which—like mass medical screenings—leads to the large number of false alarms. Face recognition is even better at another task. When you unlock your phone by looking at the screen, it performs a task called *authentication*. Unlike a perpetrator running away in the subway, you look directly into the camera, hold it close to your face, and keep perfectly still; it's virtually always you who tries to unlock your phone. This situation creates a fairly stable world: you and your phone. Errors rarely occur.

To discuss the pros and cons of face-recognition systems, one needs to distinguish between these three situations: many-to-many, one-to-many, and one-to-one. In mass screening, many people are compared with many others in a data bank; in identification, one person is compared with many others; and in authentication, one person is compared with one other. Once again, the smaller the uncertainty, as in identification as opposed to mass screening, the better the performance of the system. Recall the storming of the US Capitol in January 2021, where face-recognition systems speedily identified some of the intruders who had forced their way into the building. The general point is that AI is not good or bad but useful for some tasks and less so for others.

Last but not least, the concerns about privacy fit with this analysis. The general public is most concerned about mass surveillance by governments, not identification of perpetrators and authentication. And mass surveillance is exactly what face-recognition systems are most unreliable at doing. Understanding this crucial difference helps protect the individual freedoms valued in Western democracies against the surveillance interests of their own governments.

I Have Nothing to Hide

This phrase has become popular in discussions about social media companies that collect all of the personal data they can get their hands on. You might hear it from users who prefer to pay with their data, not with their money. And the phrase could well hold true for those of us who live uneventful lives without any serious health issues, have never made any potential enemies, and wouldn't speak up on civil rights denied by a government. Yet the issue is not about hiding or the freedom to post pictures of adorable kittens at no cost. Tech companies don't care whether or not you have something to hide. Rather, because you don't pay them for their services, they have to employ psychological tricks to get you to spend as much time as possible on their apps. You are not the customer; the customers are the advertisers who pay tech companies to grab your attention. Many of us have become glued to our smartphone, get too little sleep because of our new bed partner, find hardly any time for anything else, and eagerly await another dopamine shot via each new like. Jia Tolentino wrote in the *New Yorker* about her struggle with her mobile phone: "I carry my phone around with me as if it were an oxygen tank. I stare at it while I make breakfast and take out the recycling, ruining what I prize most about working from home—the sense of control, the relative peace."[11] Others are hurt after reading a destructive online comment from a stranger about their looks or wits. Others again drift into extremist groups that fall prey to fake news and hate speech.

The world is split between those who don't worry much about being affected by digital technology and those, like Tolentino, who believe it makes them addicted in the same way that compulsive gamblers cannot keep their minds off gambling. Yet technology, and social media in particular, could well exist without being designed to rob people of time and sleep. It is not social media per se that make some of us addicted; it is the personalized ad-based business model. The damage to its users flows from that original sin.

The Free Coffeehouse

Imagine a coffeehouse that has eliminated all competitors in town by offering free coffee, leaving you no choice but to go there to meet your friends. While you enjoy the many hours you spend there chatting with your friends, bugs and cameras wired into the tables and walls closely monitor your conversations and record whom you are sitting with. The room is also filled with salespeople who pay for your coffee and constantly interrupt you to offer their personalized products and services on sale. The customers in this coffeehouse are in effect the salespeople, not you and your friends. This is basically how platforms like Facebook function.[12]

Social media platforms could function in a healthier way if they were based on the business model of a real coffeehouse or of TV, radio, and other services where you as the customer pay for the amenities you want. In fact, in 1998, Sergey Brin and Larry Page, the young founders of Google, criticized ad-based search engines for being inherently biased toward the needs of advertisers, not consumers.[13] Yet under the pressure of venture capitalists, they soon caved in and built the most successful personalized advertisement model in existence. In this business model, your attention is the product being sold. The actual customers are the companies who place ads on the sites. The more often people see their ads, the more the advertisers pay, leading social media marketers to run experiment after experiment to maximize the time you spend on their sites and to make you want to return as quickly as possible. The urge to grab your phone while driving a car is a case in point. In short, the quintessence of the business model is to capture users' time and attention to the greatest extent possible.

To serve advertisers, tech companies collect data minute by minute on where you are, what you are doing, and what you are looking at. Based on your habits, they make a kind of avatar of you. When an advertiser places an ad, say for the latest handgun or expensive lipstick, the ad is shown to those who are most likely to click on it. Typically, advertisers pay the tech company every time a user clicks on the ad, or for every impression. Therefore, to increase the chance that you click on an ad, or just see it, everything is done to influence you to stay on the page as long as possible. Likes, notifications, and other psychological tricks work together to make you dependent—day and night. Thus, it's not your data that are being sold, it's your attention, time, and sleep.

If Google and Facebook had a fee-for-service model, none of that would be necessary. The armies of engineers and psychologists who run experiments on how to keep you glued to your smartphone could be working on more useful technological innovations. Social media companies would still have to collect specific data for improving recommendations in order to meet your specific needs, but they would no longer be motivated to collect other superfluous personal data—such as data that might indicate that you are depressed, have cancer, or are pregnant. The main motivation behind collecting these data on you—personalized advertisements—would disappear. Netflix is a good example of a company that has already implemented this fee-for-service model.[14] From the user perspective, the small disadvantage would be that we all would have to pay a few dollars every month to use social media. For the social media companies, however, the big advantage of the more lucrative pay-with-your-data plan is that the men—yes, virtually all men—at the top of the ladder are now among the wealthiest and most powerful people on earth.

Staying on Top of Technology

These examples provide a first impression of what staying on top of technology is about. Resisting the siren call to text while driving entails the ability to stay in charge and control a technology. The possibilities and limitations of face-recognition systems show us that the technology is excellent in fairly stable situations, such as unlocking your phone or for border control where your passport photo is compared with another photo taken of you. But when screening faces in real-world conditions, the AI stumbles and creates too many false alarms, which can lead to huge problems when masses of innocent people are stopped and searched. Finally, the problems caused by social media—from loss of time, sleep, and the ability to concentrate to addiction—are the fault not of social media per se but of the companies' pay-with-your-data business plan. To stamp out these severe problems, we need to go beyond new privacy settings or government regulations of online content and tackle the root of the problem, such as by changing the underlying business plan. Governments need more political courage to protect the people they represent.

One might think that helping everyone to understand the potential and risk of digital technology is a primary goal of all education systems and governments worldwide. It is not. In fact, it is not even mentioned in the Organisation for Economic Co-operation and Development's "Key Issues for Digital Transformation in the G20" of 2017 or the European Commission's 2020 "White Paper on Artificial Intelligence."[15] These programs focus on other important issues, including creating innovation hubs, digital infrastructures, and proper legislation and increasing people's trust in AI. As a consequence, most digital natives are woefully unprepared to tell facts from fakes and news from hidden ads.

Solving the problems, however, entails more than infrastructure and regulation. It requires taking time to reflect and do some serious research. Did you have to wait for a long time when calling a service hotline? It could be that your address or a prediction algorithm indicated that you are a low-value customer. Have you noticed that the first result in a Google search is not the most useful one for you? It is likely the one for which an advertiser paid the most.[16] Are you aware that your beloved smart TV may record your personal conversations in your living room or bedroom?[17]

If none of this is new to you, you might be surprised to learn that for most people it is. Few know that algorithms determine their waiting time or analyze what smart TVs record for the benefit of unnamed third parties. Studies report that about 50 percent of adult users do not understand that the marked top search entries are ads rather than the most relevant or popular results.[18] These ads are in fact marked, but over the years they have come to look more like organic search results (that is, non-ads). In 2013, Google's ads were no longer highlighted with a special background color, and a small yellow "Ad" icon was introduced instead; since 2020, the yellow color has also been removed and the word "Ad" is just in black, blending into the organic search results. Advertisers pay Google for each click on their ads, so if people mistakenly believe the first results are the most relevant, that is good for business.

As mentioned, many executives and politicians are excessively enthusiastic about big data and digitalization. Enthusiasm is not the same as understanding. Many of the overly zealous prophets do not appear to know what they are talking about. According to a study of over 400 executives in eighty large publicly listed companies, 92 percent of the executives have no recognizable or documented experience with digitalization.[19] Similarly, when Mark Zuckerberg had to testify on Facebook's latest privacy controversy to politicians from the US Senate and House, the most stunning revelation was not what he said in his rehearsed responses. It was how little US politicians seemed to know about the opaque ways in which social media companies operate.[20] When I served on the Advisory Council for Consumer Affairs at the German Ministry of Justice and Consumer Protection, we looked at how the secret algorithms of credit scoring companies are supervised by the data protection authorities, who are there to ensure that the algorithms are reliable indicators of creditworthiness that do not discriminate on the basis of gender, race, or other individual features. When the largest credit scoring company submitted its algorithm, the authorities admitted to lacking the necessary expertise in IT and statistics to evaluate it. In the end, the company itself bailed them out by selecting the experts who wrote the report, even paying their fees.[21] Ignorance appears to be the rule rather than the exception in our smart world. We need to change that quickly, not in the distant future.

Technological Paternalism

Paternalism (from the Latin word *pater*, for father) is the view that a chosen group has the right to treat others like children, who should willingly defer to that group's authority. Historically, its justification has been that the ruling group has been chosen by God, is part of an aristocracy, or owns secret knowledge or splendid wealth. Those under their authority are considered inferior because they are women, people of color, poor, or uneducated, among others. During the twentieth century, paternalism was on the retreat after the vast majority of people finally had the opportunity to learn to read and write, and after governments eventually granted both men and women freedom of speech and movement, along with the right to vote. That revolution, for which committed supporters landed in prison or gave their lives, enabled the next generations, including us, to take matters into our own hands. The twenty-first century, however, is witnessing the rise of a new paternalism by corporations that use machines to predict and manipulate our behavior, whether or not we consent. Its prophets even announce the coming of a new god, an omniscient superintelligence known as AGI (artificial general intelligence) that is said to surpass humans in all aspects of brainpower. Until its arrival, we should defer to its prophets.[22]

Technological solutionism is the belief that every societal problem is a "bug" that needs a "fix" through an algorithm. Technological paternalism is its natural consequence, government by algorithms. It doesn't need to peddle the fiction of a superintelligence; it instead expects us to accept that corporations and governments record where we are, what we are doing, and with whom, minute by minute, and also to trust that these records will make the world a better place. As Google's former CEO Eric Schmidt explains, "The goal is to enable Google users to be able to ask the question such as 'What shall I do tomorrow' and 'What job shall I take?'"[23] Quite a few popular writers instigate our awe of technological paternalism by telling stories that are, at best, economical with the truth.[24] More surprisingly, even some influential researchers see no limits to what AI can do, arguing that the human brain is merely an inferior computer and that we should replace humans with algorithms whenever possible.[25] AI will tell us what to do, and we should listen and follow. We just need to wait a bit until AI gets smarter. Oddly, *the message is never that people need to become smarter as well.*

I have written this book to enable people to gain a realistic appreciation of what AI can do and how it is used to influence us. We do not need more paternalism; we've had more than our share in the past centuries. But nor do we need technophobic panic, which is revived with every breakthrough technology. When trains were invented, doctors warned that passengers would die from suffocation.[26] When radio became widely available, the concern was that listening too much would harm children because they need repose, not jazz.[27] Instead of fright or hype, the digital world needs better-informed and healthily critical citizens who want to keep control of their lives in their own hands.

This book is not an academic introduction to AI or its subfields such as machine learning and big data analytics. Rather, it is about the human affair with AI: about trust, deception, understanding, addiction, and personal and social transformation. It is written for a general audience as a guide to navigating the challenges of a smart world, and it draws on my own research, among others, on decision-making under uncertainty at the Max Planck Institute for Human Development, a topic that continues to fascinate me. In the course of the book, my personal take on freedom and dignity is not disguised, but I do my best to stick to the evidence and let readers make up their own minds. My deeply held conviction is that we human beings are not as stupid and incapable of functioning as is often claimed—so long as we continue to remain active and make use of our brains, which have developed in the intricate course of evolution. The danger of falling for the negative AI-beats-humans narrative and passively agreeing to let authorities or machines "optimize" our lives on their terms is growing by the day, and it has particularly motivated me to write this book. As in my previous books *Gut Feelings* and *Risk Savvy, How to Stay Smart in a Smart World* is ultimately a passionate call to keep the hard-fought legacies of personal liberty and democracy alive.

Today and in the near future, we face a conflict between two systems, autocratic and democratic, not unlike the Cold War. Yet, different from that era, when nuclear technology maintained an uneasy balance between the two forces, digital technology can easily tilt the scales towards autocratic systems. We have seen this happen during the COVID-19 pandemic, when some autocratic countries successfully contained the virus with the help of strict digital surveillance systems.

I cannot deal with all aspects related to the vast field of digitalization but will provide a selection of topics explaining general principles that can be more widely applied, such as the stable-world principle and the Texas sharpshooter fallacy discussed in chapter 2, and the adapt-to-AI principle and the Russian tank fallacy discussed in chapter 4. As you may have noticed, I use the term AI in a broad sense, including any kind of algorithm that is meant to do what human intelligence does, but I will differentiate whenever necessary.

In each culture, we need to talk about the future world in which we and our children wish to live. There will be no single answer. But there is a general message that applies to all visions. Despite—or because of—technological innovation, we need to use our brains more than ever.

Let's begin with a problem dear to our heart, finding true love, and with secret algorithms that are so simple that everyone can understand them.

I   The Human Affair with AI

The problem isn't the rise of "smart" machines but the dumbing down of humanity.

—Astra Taylor

1 Is True Love Just a Click Away?

For one human being to love another: that is perhaps the most difficult of all our tasks, the ultimate, the last test and proof, the work for which all other work is but preparation.
—Rainer Maria Rilke, *Letters to a Young Poet*

Dating is just for other people to know that you're dating, people post about it all the time. Like kissing photos.
—Sophia, thirteen-year-old from New Jersey, quoted in Sales, *American Girls*

Money can't buy you love, the Beatles tell us. But can algorithms? A couple hundred dollars will buy you a six-month premium membership to online dating sites around the globe. The sites advertise secret love algorithms that may get you the perfect date. Every year, millions of hopeful customers, young and old, use online dating sites or mobile dating apps, and the trend is increasing.[1] Despite this popularity, many are unaware that algorithms are behind the selection of potential romantic partners.[2]

AI Finds You Love

An attractive young woman with long, windswept hair smiles at you from a website. Next to her is a handsome young man sporting a trim three-day beard, looking equally blissful. Close to their faces is the name of one of the largest online dating sites, Parship. Millions of singles in London, Paris, Berlin, Mexico City, Vienna, and Amsterdam embark on their quest to find true love and long-term happiness via its services.[3] Like EliteSingles, OkCupid, and a host of others, Parship is a serious agency for singles seeking a partner for life. It attracts people who hope the day will come when they won't have to date anymore. Unlike Tinder and similar dating apps, where users see little more than looks and location, Parship uses a matching algorithm based on personality and interests. Its websites and posters prominently display the same catchphrase:

Every 11 Minutes, a Single Falls in Love.

That sounds like a very attractive deal: Sign up, pay the fee, and wait for eleven minutes! Happiness is just a click away. Millions of people have signed up, hoping to be one of those who fall in love quickly.

Yet think for a moment. Every eleven minutes, a single falls in love. That would be great news if the site had only a hundred customers. But Parship has millions of customers. Let us look more closely. One person who falls in love every eleven minutes translates to about six per hour, which makes 144 per day—assuming that singles are active day and night on the website. In an entire year, that makes 52,560 enamored clients (144 x 365). It means that if the site has one million customers, only about 5 percent of the singles fall in love within a year. After searching for ten years, about half of the customers can therefore be expected to have found true love. And if the site has more than one million paying customers, the expected waiting time is even longer. In other words, there is a real chance that you will be looking (and paying) until old age; by that time, it's indeed money that bought you love. This simple check reveals a sobering truth about the success of the matching algorithm behind the persuasive slogan.

Now we know what "every eleven minutes" means. What about the second part of the catchphrase, "a single falls in love"? After all, it takes two to fall in love to make a couple. It turns out that every eleven minutes a premium member quit and, when asked why, clicked the "fell in love" button. Whether true love was in play, whether it was found online or offline, or whether it was just a handy excuse to stop paying for the service remains unknown.

Customer evaluations are consistent with this back-of-the-envelope calculation. In 1,500 evaluations of five online dating sites that Germans use, including Parship, none received an average rating of good. Only 7.7 percent said that their search was successful; the rest had quit or were still looking.[4]

In other countries and on other sites, chances appear to be on par for similar services that target highly educated singles in New York, Toronto, or Seoul. For instance, EliteSingles advertises that 381,000 new members joined per month in 2018 and that more than 1,000 singles find love through the service every month.[5] Once again, these figures sound impressive—so many happy people! But the question remains, what are your actual chances of finding love? If the figures are correct, then this amounts to roughly one in every 381 singles per month, which is about one out of thirty per year, or 3 to 4 percent. This figure is in the ballpark of the 5 percent estimated for Parship. These calculations are rough estimates based on the numbers the online services themselves provide. Just to make sure, I checked another online dating site. Jdate (for Jewish singles) reports that it has hundreds of thousands of members worldwide and that "each week, hundreds of JDaters meet their soul mates."[6] If we put these two numbers together, the result is about one soul mate among 1,000 singles per week, which translates into some fifty-two in 1,000 per year and once again a chance of about 5 percent per year. This long waiting time fits well with one of the site's advertised "success stories": the case of Ryan, who spent fifteen years scrolling and clicking on faces until he finally found his soul mate through the site.[7]

Love algorithms can deliver three kinds of services: access, communication, and finding true love. Access enables customers to meet potential partners whom they are otherwise unlikely to meet. It is particularly helpful for people who live socially or physically isolated lives or do not meet current social norms, such as those with disabilities or strict religious beliefs. Communication via computer-mediated interaction before meeting face-to-face is another service. Together, access and communication are the true feat of online dating. But increased access also comes with a downside when the chances of finding true love are slim, as we have just seen. After experiencing twenty-two meh dates instead of a single one that went wrong, you might well feel worse with that amount of misfortune and choice overload. A review of studies on online dating concluded that "no compelling evidence supports matching sites' claims that mathematical algorithms work—that they foster romantic outcomes that are superior to those fostered by other means of pairing partners."[8]

To check these results, I contacted major reputable online-dating agencies. After all, several dating agencies—including eHarmony, perfectMatch

.com, and Chemistry.com—claim to have powerful "scientific" algorithms, even though none of them have been shown to be reliable and valid by standard scientific methods.[9] When I inquired about the actual number of paying customers, the success rate, and how they determined it, I received friendly but firm refusals to answer these questions.[10]

Finally, let us look beyond online-dating agencies. Do couples who met online—via social networks, chat rooms, dating agencies, or others—break up less often, and are they more satisfied with their relationship than those who met offline? A classic representative study with over 19,000 married US residents found that those who met online had fewer marital break-ups than those who met offline and also reported slightly higher satisfaction with their marriages. Note that those who met through online-dating agencies using love algorithms were not more satisfied than those who met on other online sites.[11] On the other hand, more recent studies in the US reported that breakup rates (both marital and nonmarital relationships) were higher for couples who met online, and representative studies in Germany and Switzerland reported no difference in couples' satisfaction with their relationships whether they met online or offline.[12] While these results are inconsistent, several studies consistently report that the proportion of couples who met online is higher among same-sex couples and that online dating brings together people from more diverse backgrounds in education, race, and ethnicity, which can still be barriers offline.[13] This increase in diversity, however, appears to be largely due to the fact that couples who met online tend to be younger.

Thus, the jury is still out on whether online search leads to more satisfaction and fewer breakups than offline search. All in all, unless you belong to the lucky 5 percent, you might as well spend your time and money on meeting colleagues after work, going to parties, traveling, walking a dog, or participating in a local online community that shares one of your personal passions—that is likely to be a faster route to happiness. Cupid's arrow might strike in unexpected places.

How Do Love Algorithms Work?

Love algorithms are big secrets. Like algorithms for credit scoring, predictive policing, and page ranking, they are proprietary. And every agency uses different algorithms. That makes it difficult to find out exactly how they

Feature	Adam	Eve
Wants children	Yes	Yes
Center of attention	Yes	No
Likes to cook	No	No

Figure 1.1
Two profiles with three features.

work. Nevertheless, we know the basic procedure, which is one of the simplest versions of an algorithm. On serious dating sites, customers fill out a survey about their values, interests, and personality. A survey may consist of more than a hundred questions. The answers are then transformed into a customer profile.[14] To make it simple, consider two profiles with only three features (figure 1.1). Adam wants children, enjoys being the center of attention, and does not like to cook. Eve also wants children, does not enjoy being the center of attention, and also does not like to cook.

The first principle on which the match between Adam and Eve is calculated is *similarity*. In the case of wanting children, similarity is crucial. The simplest algorithm would just count the number of times both agree, which would be two out of three, or two-thirds. But looking at similarity alone wouldn't work. What matters is not only similarity but also *complementarity*. For instance, Adam is keen to be the center of attention. He might not be thrilled by a partner sharing that desire, with the prospect of eternally fighting over who gets to be in the limelight. A partner with complementary interests, that is, someone who dislikes being the center of attention, is a better fit for Adam. In the same way, if both hate cooking, they are similar but likely not the best match. To find out whether a customer wants a similar or a complementary partner, some agencies ask them in the survey what they expect from their ideal partner.

Finally, there is a third principle, *importance*. Not all features are equally important. Some dating services estimate the importance themselves, while others ask their customers to indicate how important an attribute is for them: irrelevant, a little important, and so on. The algorithm, however, needs numbers as inputs instead of verbal answers. The dating app OkCupid, one of the few that has made their algorithm fairly transparent, uses irrelevant = 0, a little important = 1, somewhat important = 10, very

Features	Adam's ideal partner	Eve	Importance for Adam	Points
Wants children	Yes	Yes	10	10/10
Center of attention	No	No	50	50/50
Likes to cook	Yes	No	1	0/1
				Score: 60/61

Figure 1.2
Adam's point of view. How satisfying is Eve for Adam? The answer is 60 out of 61 possible points, or 98 percent, which is extremely good.

Features	Eve's ideal partner	Adam	Importance for Eve	Points
Wants children	Yes	Yes	250	250/250
Center of attention	No	Yes	50	0/50
Likes to cook	Yes	No	10	0/10
				Score: 250/310

Figure 1.3
Eve's point of view. How satisfying is Adam for Eve? The answer is 250 out of 310 possible points, or 81 percent, which is not as good.

important = 50, and mandatory = 250.[15] Now we can put all three principles together and determine how satisfactory Eve is for Adam (figure 1.2).

On the first two attributes, Eve matches Adam's ideal partner. Because the importance he sets on these is 10 and 50, respectively, the result is 60 out of 60 points. The fact that Eve doesn't enjoy cooking is of little importance to Adam, so she gets 0 out of 1 point for the third attribute. In total, Eve gets 60 out of 61 points, which is about 98 percent.

In the same way, the algorithm calculates how satisfying Adam is for Eve (figure 1.3). Adam wants children and so does Eve's ideal partner. Because children are mandatory for Eve, this similarity results in 250 out of 250 points. Yet there is no match for the other two attributes. The algorithm calculates a total of 250 out of 310 points, which is 81 percent.

In a last step, the algorithm calculates the total match between Adam and Eve by taking an average between their scores, which would be 89.5 percent in our example.[16]

The reality of profile matching is more complicated than this, but the basic logic remains the same. There are more features, and the features may be quantitative, such as age and income. Similarity tends to count for hobbies and values, while complementarity is surprisingly often desired when it comes to education and age, especially for heterosexual couples. Women who subscribe to dating services in Boston, Chicago, New York, and Seattle desire men with a higher level of education: the higher the better. But men in those same cities do not want the most educated women. On average, they prefer women with undergraduate degrees and find anyone above that level, with a master's or PhD, less attractive. Similarly, the average woman's desirability peaks at age eighteen, while men's desirability peaks around age fifty.[17] This striking discrepancy was also observed among OkCupid's customers. Women find men most attractive if they are about as old as they are, give or take two to three years. The average man, however, always prefers women in their early twenties, no matter his age.[18] These men never grow up. Their preferences follow a basic insight of evolutionary psychology: men tend to be attracted by cues that indicate high fertility, such as youth and smooth skin, while women tend to be more interested in cues that indicate a man's ability to support a family, such as wealth and a good education.

All in all, it is not that difficult to understand the basics of algorithms that match two profiles, even if they are guarded as top secret. In general, an algorithm transforms input numbers into output numbers, such as profiles into probabilities of matches.

A Profile Is Not the Person

If love algorithms can use profiles, similarity, complementarity, and ratings of importance, why can't they quickly find the ideal partner for life? Understanding these four principles helps acquire a realistic understanding of what these algorithms can and cannot do. Let us begin with the profiles. In a face-to-face meeting, the "data" are rich and complex: a smile and a gesture, the humor reflected in someone's eyes, the tone of voice, the way the other person asks questions, the intensity or superficiality with which a person listens. And then there are touch and scent, which can be crucial to mating compatibility, especially for women.[19] Profiles, in contrast, are based not on real interactions but on the responses given in the initial survey. A profile is not a person, it is a self-presentation, not necessarily of true interests and values. Even when "personality traits" are used in the profile, the algorithm infers these from self-reports. For instance, some sites ask whether attributes such as "sexy" and "plain" apply to you, and how "rational," "opinionated," and "selfish" you are. What would you say? Few people are realistic and candid when yearning for the perfect date. When you are asked for your interests, would you admit to being a couch potato with no special interests? Or acknowledge that you are a first-rate dancer, which might intimidate most candidates? As a consequence, self-reported interests and personality predict romantic love only weakly.[20] The same result is known from speed dating.[21] What people say they prefer does not match with the actual choices they make.[22]

Now consider the principles of similarity and complementarity. There is a saying that birds of a feather flock together, and yet another that opposites attract. Whatever the truth, love algorithms can compare similarity and complementarity only by looking at people's profiles, not their actual behavior. And this turns out to be a crucial limitation. A review of 313 laboratory and field studies on the topic showed something striking: if people did not know each other, higher similarity between their profiles (attitudes and personality traits) clearly went along with higher mutual attraction. After a short interaction of only a few minutes or hours, however, that attraction largely faded away when people met face-to-face.[23] This means that similar profiles do spark initial attraction but appear to be irrelevant for a real relationship. It may also explain why people find partners with a highly matching percentage attractive, and why at the same time the match percentage contributes little to finding true love. In fact, a study of over 23,000 married persons in Australia, Germany, and the United Kingdom found that the degree to which partners were satisfied with their relationship had very little to do with how closely their personality trait profiles resembled each other.[24] Instead, a person's own personality—such as being agreeable, conscientious, and emotionally stable—was related to how satisfied they were with a relationship. Some individuals will mess up any relationship, no matter how suitable their partner is. Online dating sites such as eHarmony have recognized this finding and reject customers who do not appear to be emotionally healthy.[25]

Finally, consider the principle of importance. You may have wondered why OkCupid uses the numbers 1 to 250. As mentioned above, the ultimate reason is that algorithms need numbers to calculate matching percentages.

Like other algorithms, love algorithms take numbers as inputs and transform them into output numbers. Dating services either intuitively assign numbers to their customers' importance statements, as OkCupid does, or try to estimate them from data. For the latter, the service would need reliable data about which combinations of profiles led to lasting love. I am not aware of any service that bothers to follow up with their customers and systematically collect these statistics. Fixed numbers also assume that customers' ratings of importance are stable. However, some judgments may well change in the course of a relationship, such as when a partner learns to appreciate the other's interests and values that were considered of little importance beforehand. Whether these numbers capture the true importance therefore remains uncertain.

All in all, the big feat of dating platforms is access, not the ability to find true love. That may come when people meet face-to-face. Self-presentation in profiles and the principles of similarity, complementarity, and importance calculated from these profiles are the basics of matching algorithms but not necessarily the main ingredients for a successful relationship. This can explain why love algorithms often flop. But as we will see in the next chapter, a more general reason exists: unlike chess, finding true love is a game riddled with uncertainty, and that is where algorithms run into problems.

Courtship Adapts to Software

It's easy to think of algorithms as "neutral" tools to speed up access to the best romantic partners. Yet they are not. As algorithms increasingly populate our world, they have the power to influence our values—even when they don't work very well. That power extends to courtship as well. For instance, a blind date is hardly possible today; each party will look the other one up on social media. Like other technologies, algorithms change behavior and ultimately our desire. To begin with, software can turn courtship into an optimal search problem and decrease commitment.

Always Looking for Someone Better

As one story goes, a young woman has found the man of her dreams, and they are together in bed. When he leaves for a minute to go to the bathroom, she automatically reaches for her smartphone, opens Tinder, and begins browsing and swiping for other men. Surprised at what she is doing, she cannot really explain it. She just can't stop. Stories like this illustrate how software can take over and control behavior.

Easy access to numerous partners can turn a happy "satisficer" into a restless "optimizer." *Satisficing* is a term from decision theory that, in the context of dating, means choosing a partner who is satisfactory. To do so, you need to develop a desirable standard (an *aspiration level*), pick the first partner who meets it, and stop looking any further. Stopping search is a prerequisite for committing to a happy long-term relationship. Optimizing, in contrast, means looking for the absolutely best partner and nothing less. Since there are plenty of fish in the sea and it is impossible to know what other alternatives might pop up, optimizing leads to disenchantment and a culture of "always looking for someone better." Even after accidently stumbling across the very best partner (if such a person exists), an optimizer would not know it and would continue searching for a better one, openly or secretly. Dating sites that do not limit the number of viewable candidates per day prompt this behavior. A satisficer, by contrast, would cancel the online dating service immediately and embark on a serious relationship.

Yet it is in the very interest of dating agencies that customers stay on board as long as possible. They would make little profit if everyone found true love in eleven minutes and quit. That is why many agencies emphasize providing access over brokering true love. Access to large numbers of potential partners is one way to keep users on the app, just as supermarkets offer a hundred varieties of mustard and jams to keep customers looking and buying. Accordingly, many users design their dating profiles as if these were product descriptions in online shopping and dispose of other products with a swipe. In their world, true love means searching nonstop for the best deal.

Optimize Your Profile, Not Your Person

The epigraph to this chapter cites Sophia, a thirteen-year-old from New Jersey, who gave her own view on dating. Many girls like Sophia spend large amounts of their time competing on social media and, later, on the online sexual marketplace, managing their selfies and reputation. "I feel like I'm brainwashed into wanting likes," she explained.[26] Self-presentation is of course nothing new, but digital media offer handy tools to edit one's self. The psychologist Robert Epstein described a date he had with a woman he met online.[27] When they finally met for coffee, the woman did not look anything like the woman on the pictures she had posted. She was in marketing and thought it a good strategy to post photos that attract "customers." Later, Epstein noticed that she'd replaced the photos with those of yet another woman.

The lifeblood of advertising is appearance and duplicity to attract attention. In their online profiles, people lie about their age, marital status, income, height, and weight. Researchers from Cornell measured and weighed people, and then checked the numbers against their online profiles. On average, people were one inch shorter and five pounds heavier than what they'd posted.[28] And the shorter and heavier people were, the bigger the discrepancies. A study of over 5,000 subscribers to online dating services in Boston and San Diego found that women in their twenties listed their weight on average as five pounds lower than the corresponding weight in the general population, which increased to seventeen pounds for women in their thirties and to nineteen pounds for women in their forties.[29] No such discrepancies in weight were found for men. Women also prefer to list their age as twenty-nine, thirty-five, and forty-four more often than other ages.[30] Men, in contrast, tend to overstate their income, and there is a reason why. When men claimed to have an income of $250,000 rather than less than $50,000, three times as many women replied. Men also lie about their marital status: likely one out of eight male online suitors is married.[31] The researchers who measured and weighted people used hard facts to determine the extent of cheating. But even when asked in surveys, more than half of US participants openly admitted to lying on their dating profile, a higher number than that of their British counterparts.[32]

Some online dating services appear to resort to tricks themselves. They claim to have more customers than in reality and maintain that these are highly satisfied.[33] When asked about the actual numbers, the agencies become highly secretive.

Female Bots

Have you ever been contacted by a distraught widow in Somalia seeking to transfer her multimillion heritage to your safe bank account? If you reply, she might soon ask you to send money to cover customs and taxes in return of a promised $10 million. Did you ever receive an email that you are the lucky winner of $250,000 in an overseas lottery? One might believe that nobody with a brain would fall for such scams. Yet some people are taken in by hope. As one British victim explained: "That amount of money gives you dreams, and you don't want them taken from you."[34] Finding potential victims was once a trial-and-error operation, but the AI tools behind Facebook have made it easier. As they do for every advertiser, Facebook's machine-learning algorithms help identify those users who are most likely to click on the fraudulent ad. Companies behind such scams, after camouflaging their ad content, pay Facebook on average $44,000 in advertising fees and make a return of $79,000 from the victims.[35] In the UK alone, around one million adults fall victim to mass-marketed scams per year and together lose about £3.5 billion.[36]

Romance Scams

Still, most people resist these temptations because the messages are unsolicited and from strangers. To widen their circle of victims, international criminal groups therefore search for people on dating websites, which are designed for connecting strangers. They set up false user profiles, typically with edited pictures of stolen photographs, preferably army officers, engineers, and attractive models. Then the crooks start a romantic relationship online, declare their love for the victim, and ask that they move from the dating site to instant messaging or email to start an exclusive relationship. During the grooming stage, many a victim falls in love.[37] Some weeks or months later, after trust and love are gained, the criminals first ask for small amounts of money, and then for more. The plot may involve other criminals in the group who pretend to be doctors and inform the victim that their loved one has been rushed to a hospital, requesting money to pay the hospital bills. The game ends when the victim finds out about the scam or has no money left. According to the FBI, victims lose an average of $14,000.[38] Online dating scams are now commonplace, and many victims are too embarrassed to go public. But there is an easy way to detect a scam. If your online sweetheart begins to ask for money, for whatever reason, you can guess why. If you do an image search and the person's photo appears under other names, you know. There is a simple rule to avoid heartbreak and an empty bank account: never send money to someone you met online and don't know in person.

Besides the financial loss, victims experience the loss of a relationship. For most, the loss of the object of love is more upsetting. As one victim explained, "I'd been talking to this guy for so long and how could one day you get up and be told it's a scam, how can that go away? I mean your feelings don't change." Many are embarrassed and shocked to learn that they had fallen prey to a scam: "Well you are mentally raped. Because they've totally picked your brains and everything else," and "God, how can I be so stupid!"[39]

The high-tech versions of these human swindlers are bots. Bots are autonomous software agents that make decisions without human intervention. Unlike benevolent bots, such as editing bots on Wikipedia and web crawlers, which are designed to support human users, malevolent bots are designed to exploit us. Consider the Canadian online dating service Ashley Madison, a cheater dating site that was founded with the slogan "Life is short. Have an affair." Not surprisingly, there was a shortage of female customers. The business model was that male customers had to pay a fee to initiate a conversation with a female. When hackers stole all of the site's customer data—including emails, names, addresses, and sexual fantasies—it turned out that Ashley Madison had "employed" over 70,000 "female" bots to send over twenty million fake messages to male cheaters.[40] The hackers threatened to leak this data if Ashley Madison did not immediately shut down its services. When the infidelity dating site refused, the hackers released the personal data of thirty-two million customers, including those who had paid a $19 fee to be deleted, which was yet another fake service. "This will wreck my marriage," said a Kentucky man who had used the site for years to cheat on his wife.[41] Others were angry about their exposure after only receiving fake profiles of potential partners. And some 1,200 of the email addresses ended with .sa, indicating customers from Saudi Arabia, where adultery can mean the death penalty.

You Caught His Eye

The most unexpected part of this sad story is that dating sites themselves take advantage of scammers to lure customers into purchasing subscriptions. In the US, the Federal Trade Commission (FTC) sued the online service Match for using fake love interest advertisements to trick hundreds of thousands of consumers.[42] Match owns Match.com, Tinder, OkCupid, PlentyOfFish, and other dating sites.

Suppose you are looking for "the one." Most online services offer to create a profile for you free of charge, in addition to giving you the option of
purchasing a subscription. So you might just sign up because it's free and
see what happens. To your surprise, your profile quickly attracts attention,
and the dating service reports that someone has expressed interest in you
(figure 1.4). But, because you didn't subscribe, you cannot respond. That
persuades you to purchase a subscription and write back to the person.
Too late; the profile is now "unavailable." Oh no, you waited too long and
should have subscribed directly in the first place. Now you may well have
missed "the one."

But don't worry, it's not what you think. Millions of the interested
"eyes" are scammers who want you to become their romance scam victim.

match.com

He just emailed you!

You caught his eye and now he's expressed

interest in you... Could he be the one?

READ HIS EMAIL >>

You will be notified when other Match.com members express interest in you.

Please note: This email may contain advertisements.

Match.com P.O. Box 25472, Dallas TX 75225

Figure 1.4
"You caught his eye." A message sent to people who have put up a profile on a dating
service for free in order to lure them into a paid subscription. Only by paying can
they respond to the person who expressed interest. The Federal Trade Commission
sued the dating service Match, which owns Match.com, Tinder, PlentyOfFish, and
other dating sites, because Match knew that many of these contacts were scammers,
having flagged their accounts as fraudulent. It nevertheless forwarded these contacts,
but only to their nonsubscribing customers to nudge them into subscribing. Source:
FTC, "FTC Sues Owner of Online Dating Service."

According to the FTC, the dating service was well aware of the scammers and had flagged their accounts as fraudulent, yet nevertheless forwarded their messages to make their nonpaying customers curious and subscribe. In this way, the dating services exploited the scammers to get subscriptions, deceiving their own customers and profiting from fraud. And the ploy works. The FTC reported that over two years, about 500,000 curious customers purchased a subscription within twenty-four hours after a receiving a fraudulent message such as "You caught his eye." Those who received the disappointing response "unavailable" were actually the fortunate ones. Match does try to prevent contact with scammers, but only for their subscribers. For the nonsubscribers, they used scammers as bait. That may explain why customers are often puzzled that the number of notifications of interest faded away the moment they subscribed.

Arranged Marriages Go Online

In contemporary Western societies, dating happens between two individuals. Marriage usually comes last. That is not so in all cultures. In the traditional Indian system, parents decide on the choice set. Marriage comes first, then sex, then love.

The Indian Institute of Management in Bangalore, with its landscaped gardens, provides an idyllic environment for learning. Its motto is "Let our study be enlightening." While teaching there at a winter institute on decision-making, I was approached by Sanjey, an assistant professor. He told me his story about finding true love. In the traditional Indian system, the family of a prospective bride or groom searches for a match. In the digital age, marriage advertisements are typically placed in the "wanted" section of matrimonial websites, not unlike in the West, but it is often the family who advertises, not the individual. The ads typically provide the age and height of the young man or woman, and some of the following: their profession and salary, the expected salary of the partner, their father's profession, their caste, and whether a dowry is required. From the responses on the website, the family may select three to six possible partners. Then the parents pick one of the candidates and arrange a meeting at a restaurant, where the young people meet for the first time and have a chance to talk to one another—in the presence of their parents. After returning home, the parents ask their children whether they would accept the candidate as a partner for life. If both young people accept, the wedding preparations go ahead. If one of them rejects the other, then the process is repeated with the next candidate on the list.

Sanjey disliked this procedure. But not, as one might think, because he wanted to do it the Western way and have more choice. Rather, he did not want to have any choice at all. He asked his parents to decide for him, trusting their experience, and simply hoped that they would select an attractive, lovable wife. When I asked him why he did not want to see at least a few of the women, he explained that he did not want to hurt a woman's feelings by rejecting her. That was a remarkable statement, so unlike the spirit of casually rejecting candidates online literally with a swipe of a hand. Sanjey married the woman his parents selected, and both of them told me they had since fallen in love and have been happy together for years. In the world of Tinder, access is an asset. In Sanjey's world, access implies rejecting and hurting others' feelings.

Trusting one's parents and their experience is unthinkable, even irrational, for most Westerners. But as Sanjey and other researchers at the Indian Institute of Management politely pointed out, why would Americans and Europeans consider it rational to seek a partner by randomly meeting someone in a disco, where they can hardly hear each other and have had a drink too many? Will parents' choices or love marriages be the future? There is a third option. The increasing faith in the omnipotence of AI may well put an end to both systems.

The Ultimate Convenience: Let the AI Arrange Our Marriages

We dream of autonomous cars because they are purportedly safer than human drivers. Why not autonomous partner algorithms? The argument in favor of both is similar. Almost all deadly car accidents are due to human failure: drunk driving, drugs, fatigue, or distraction by one's cell phone. Self-driving cars would protect people from their own errors. Similarly, half of all first marriages and two-thirds of second marriages in the United States end in divorce, leading to regret for having married the wrong partner as well as to emotional issues for the children. Why not put an algorithm "in the driver's seat" that protects people from making bad decisions? An autonomous AI would match the right people before they go astray.

One philosopher of AI advises: "We should defer to the superintelligence's opinion whenever feasible."[43] In this vision of rational choice, you trust an algorithm, not yourself or your family, to choose your romantic partner. Taken to the extreme, the result would be an autonomous AI that arranges marriages without any human interference. All you have to do is to turn up on time at the location you are told.

Such an AI is science fiction, but the science fiction exists. "Hang the DJ," an episode of the British TV series *Black Mirror*, named after the dark display of a smartphone, projects us into such a future world where romantic relationships are no longer individual decisions. In this world, a love algorithm matches people into relationships, and everyone follows the algorithm's choice without hesitation. Moreover, each relationship comes with an expiration date, upon which the lovers dutifully separate. After separation, the algorithm pairs each individual with a new partner until the next expiration date.

Today we are already told that algorithms can predict personality better than humans can.[44] Wouldn't it be foolish not to follow the AI? For some, to defer to its selection may be the ultimate convenience: no flirting necessary, no rejection possible, no feelings ever hurt. Everything is optimized, and no time and money are wasted on finding a partner on their own or through dating agencies. For others like me, that vision is the final romantic nightmare. The protagonists in the *Black Mirror* episode, where everyone believes and obeys the matching AI, are a couple who take their lives back into their own hands. They fall in love and decide, against the will of the algorithm, to stay together. But this romantic couple is the odd one out.

2   What AI Is Best At: The Stable-World Principle

If one could devise a successful chess machine, one would seem to have penetrated to the core of the human intellectual endeavor.
—Allan Newell, J. C. Shaw, and Herbert A. Simon, "Chess-Playing Programs"

Imagine how much harder physics would be if electrons had feelings.
—Richard Feynman[1]

Why can AI win at chess but not find the best mate? After all, the goal appears to be similar: to assign a score to each move or candidate, and then pick the best one. Chess algorithms such as Deep Blue assign scores to billions of possible positions it can foresee, just as love algorithms assign matching scores to millions of potential partners. It works out marvelously for chess. Why not for everything else?

Herbert Simon was one of the founders of artificial intelligence, and the only recipient to date of both the Nobel Memorial Prize in Economic Sciences and the Turing Award, which has been dubbed "the Nobel Prize in computing." Simon firmly believed that once a machine can beat the best chess player, it has reached the essence of human intelligence. In 1965, he predicted that within twenty years, machines would be capable of doing any work that humans can do.[2] The belief that chess is the pinnacle of the human intellect was so self-evident (and still is to some) that it was shared by both early enthusiasts such as Simon and critics of AI as well. For instance, when the philosopher Hubert Dreyfus tempered Simon's enthusiasm in his widely read book *What Computers Can't Do* of 1979, he nevertheless referred to chess as the core of general intelligence, only to point out that computers have been unable to win against humans.[3]

When, finally, Deep Blue beat chess world champion Garry Kasparov in 1997, AI therefore appeared to be far along the path to acquiring human-like intelligence. All that is missing appears to be more computational power and more data for AI to outsmart us in every respect. And computational power is no longer a scarce resource. According to Moore's law, computing power—the number of transistors in integrated circuits—doubles every two years or so. This exponential increase was indeed essential for the victories of AI in chess and Go. Simon's equation of this success with reaching human intelligence underlies the AI-beats-humans argument in the introduction to this book. Following Simon, popular writers now argue that it will not take long until we succeed in building an awe-inspiring superintelligence that will exceed us in everything we know and everything we do.

I am a great admirer of Simon's work, but here he overlooked a fundamental issue, which others have also ignored.

The Stable-World Principle

There is a crucial difference between games such as chess and problems such as finding a romantic partner. In chess, each position can be represented by a profile that specifies the location of each figure, from pawn to king. A chess computer does not need to make an inference about what the true position is because the profile *is* the position. There is no uncertainty. In many other situations, such as in online dating, uncertainty abounds. Although each person has a profile, as we have seen, the profile is not the person. People tend to edit their profiles, and even when they do it scrupulously, a profile does not capture the rich dimensions of a human being.

This insight is expressed more generally in the *stable-world principle*:[4]

> *Complex algorithms work best in well-defined, stable situations where large amounts of data are available. Human intelligence has evolved to deal with uncertainty, independent of whether big or small data are available.*

The rules of chess and Go are well-defined and are stable in the sense that they hold today and tomorrow. There is no uncertainty about the nature of the rules and whether they may unexpectedly change in the future. Between romantic partners, in contrast, the rules of conduct need to be negotiated and can be violated.

The stable-world principle applies to forecasting the future as well. To successfully predict the future, one needs a good theory, reliable data, and a stable world. In August 2004, NASA launched the probe MESSENGER, which entered its orbit of the planet Mercury in March 2011, at exactly the location NASA had predicted more than six years beforehand. This incredible feat was possible because a good theory of movements of the planets exists, astronomical data are highly reliable, and the movement of Mercury remains stable over time, without being significantly influenced by human behavior. Accordingly, AI is superb at dealing with such stable situations, from using face recognition for unlocking your phone to finding the fastest route to your destination to sorting and analyzing large data in accounting.

But tech companies often try to predict human behavior without a good theory, reliable data, or a stable world. If you apply for a job, an algorithm may screen your application and recommend whether you should be invited for an interview. If you get arrested, the judge may consult a risk assessment tool to calculate the probability that you will reoffend before your court date, and then decide whether you should be bailed or jailed. If you get cancer, the hospital may rely on a big data algorithm to design personalized treatment for you. If you are a social worker, you might be sent to families in your community whom an algorithm deems to be at the highest risk. In all of these situations, there is a lack of good theory, reliable data, or a stable world. Hence, AI's miraculous power is often a mirage.

The distinction I make corresponds to that between *risk* and *uncertainty* originally made by the economist Frank Knight. In situations of risk, such as roulette, we know all possible outcomes (the numbers 0 to 36) in advance, along with their consequences and probabilities. In situations of uncertainty, in contrast, we cannot know all possible outcomes or their consequences ahead. That is the case when hiring an employee, forecasting an election, or predicting infection rates of the flu or COVID-19. Finance experts use the terms *radical uncertainty* and *black swans* to characterize a world where the future is unknowable and surprises happen.[5] In these situations, Knight argued, calculation is not enough; instead we need judgment, intelligence, intuition, and the courage to make decisions. Many situations are a mixture of risk and uncertainty, which means that both machine calculation and human intelligence have a role to play.

The stable-world principle clarifies that with increasing computing power, machines will soon solve all problems in *stable* situations better than humans. For instance, we may see a program that outperforms humans in all games with well-defined rules. Yet the same does not hold for unstable situations. If the future is unlike the past, collecting and analyzing big data—which is always from the past—can lead to false conclusions. With this insight, we can better understand where complex algorithms fed with big data are likely to be successful and where humans are indispensable.

Successes of AI

"I, for one, welcome our new computer overlords," said Ken Jennings as he acknowledged his defeat in the quiz show *Jeopardy!* by IBM's supercomputer Watson in 2011.[6] The new overlord was room-size and came with twenty tons of air-conditioning equipment. Watson is named after IBM's founder, Thomas J. Watson. It has a DeepQA (question answering) algorithm that was trained on thousands of question-and-answer pairs from the show to learn which answer matched with which question. There is no question that Watson's performance in the game show was truly impressive.

The stunning victories of AI over human experts to date have all been in well-defined games, from checkers to backgammon to Scrabble. *Jeopardy!* already has strict rules, but for Watson to win, even these had to be adapted by excluding certain types of questions.[7] In May 2017, the computer program AlphaGo won against the world's top-ranked Go player at the time, Ke Jie. The games drew 280 million viewers in China, where Go champions are national heroes with the status of rock stars.[8] Like chess, Go is a well-defined game with fixed rules that cannot be negotiated by the players. In December 2017, AlphaGo's descendant, AlphaZero, was introduced and beat its predecessor. Both use deep neural networks, whose algorithms are not designed by humans but learned by machines. AlphaGo learns from games played by human experts, while all that AlphaZero needs to know are the rules of the game, without any further human input. It just uses computing power to play millions of games against itself and learn how to win by trial and error.

AlphaZero also beat the best human players at chess and shogi (Japanese chess). Yet it would be an error to believe that AlphaZero is capable of doing everything. It is specialized to play games for two players that have known and unchangeable rules. It does not work for driving a car, teaching children, finding true love, taking over the world, or other practical matters fraught with uncertainty.[9] Similarly, the recommender algorithm in Google's search engine is highly specialized and cannot play Go. In fact, AlphaGo and Google's search engine have very little in common.

Another example of AI successes are face-recognition systems used not only for border controls and unlocking phones but also for identifying one's friends in photos on online social networks, verifying identities at ATMs, and checking into hotels. In an experiment, Google's automated face-recognition system was fed huge amounts of pictures, using more than 1,000 hours on a network of powerful computers. When it was then given the task of determining whether two photographs show the same person, it was accurate in 99.6 percent of the cases, which is the level at which humans perform the same task.[10] These systems work best if you hold your face still in front of the border control camera or your smartphone— that is, in controlled authentification tasks, where one face is compared with one photo. As noted in the introduction, face-recognition systems perform substantially less well in mass screening—that is, in less-controlled situations, where many individuals are compared with many other individuals. For instance, when the Cardiff, England, police screened the faces of 170,000 soccer fans at a game in 2017, the system reported 2,470 matches with a criminal database of half a million images, 2,297 (93 percent) of which were false alarms.[11] Similarly, when Amazon's facial recognition system compared photos of the 535 members of the US Congress with a criminal database, it reported twenty-eight members that matched, all falsely.

As a final example, consider fraud control. One of the reasons why you may pay too much for health insurance is a form of fraud practiced by corrupt doctors and pharmacies. Here is how it works. If a drug costs $100 and the reimbursement rate is 90 percent, then the pharmacy gets $90 back from the insurer. If the pharmacy presents prescriptions written by a doctor without actually having sold the medication, it can make illegal profit. In Portugal, for instance, one doctor wrote 32,000 prescriptions for expensive drugs in one year—which boils down to one fake prescription every three minutes.[12] The prescriptions bore the names of deceased patients or the faked signatures of deceased doctors. These doctors and pharmacies accounted for some 40 percent of all public expenditure fraud in the country. To put a stop to this, the Portuguese National Health Service introduced an electronic prescription program requiring doctors to complete the prescription and send it to their patients by text message or email, and reported that the system could reduce fraud by 80 percent. In this case, the surveillance potential of software has been clearly used for the better of the health system.

There are countless examples of AI surpassing human capability, from industry robots that tirelessly and precisely repeat the same movements to search engines that can find words and phrases in large bodies of text. In general, my thesis is that the better defined and more stable a situation is, the more likely it is that machine learning will outperform humans.

The moment human behavior enters the field, uncertainty appears and predictions get difficult. AI can get into trouble if problems are not well-defined or situations are unstable, or both. That holds not only for the success of romantic relations but also for anticipating the next large financial crisis, which we may likely miss as we did the crisis of 2008.

The distinction between stable, well-defined problems and unstable, ill-defined ones is reminiscent of that between "known unknowns" and "unknown unknowns," in the NASA terminology popularized by former US Secretary of Defense Donald Rumsfeld. Yet the distinction is not carved in stone: most situations are a mix of both. For instance, translating from one language into another is regulated by a stable set of grammatical rules but also involves mastering ambiguous terms, phrases with multiple meanings, irony, and other sources of uncertainty.

Psychological AI

In 1957, Herbert Simon predicted that within ten years, a computer would beat the world's chess champion.[13] And he was right, apart from the timing. But that victory did not happen in the way Simon imagined.

For Simon, AI meant *teaching computers how human experts solve a problem*. The computer is the student, the human the teacher. Researchers extract the heuristics (rules of thumb) of strategic thinking used by experts and program these into a computer, which can process the rules faster and without errors. Let's call this teacher-student vision *psychological AI*. It is the original meaning of AI, where the *I* refers to the intelligence of humans mimicked by a machine. Simon and his students went a long way to extract the rules by observing chess masters and asking them to think aloud while playing. It worked for some problems, yet never succeeded in beating a world champion in chess.

The program that beat Kasparov in 1997, IBM's Deep Blue, was not based on the idea that human intelligence and machine intelligence are two sides of the same coin. Instead, it relied on massive computing power to examine 200 million positions per second. The machine calculated: If I do A, and he does B, then I do C, and he does D, and so on, where will I end up? Kasparov, by contrast, could evaluate perhaps three positions per second. When Joe Hoane, one of Deep Blue's programmers, was asked how much of his work was devoted specifically to artificial intelligence in emulating human thought, he responded dismissively: "It is not an artificial intelligence project in any way. It is a project in—we play chess through sheer speed of calculation and we just shift through the possibilities and we just pick one line."[14]

Simon tried to build a machine in the image of a human. IBM did not. Nor did Google when building AlphaZero. Their engineers relied on *machine learning*, which is another branch of AI that includes deep neural networks and other algorithms, without trying to mimic human intelligence. The *I* in machine learning AI has no relation to intelligence as we know it, which is why the term *automated decision-making (ADM)* is often used instead. Psychological AI turned out to be a bust for chess, while machine learning using brute-force calculation succeeded. That success was also perceived as relinquishing the dream of building an artificial intelligence that resembles human intelligence. And there is a fundamental difference between human intelligence and machine learning. The chess program using brute force doesn't know it is outsmarting a player. In fact, it doesn't even know it is playing something called chess; it is simply good at it. Brute computational power is high-speed calculation, not intelligence.

Is psychological AI a bad idea? Not at all. It has its place, although not, as Simon believed, in games like chess and other well-defined situations. The stable-world principle provides a different perspective: Psychological AI is likely to be successful instead in situations of uncertainty, such as predicting the future. After all, humans evolved heuristics to deal with uncertainty, and psychological AI aims to program these heuristics into a computer.

It is interesting that Simon was also one of the fathers of research on heuristics and is known for the insight that, under uncertainty, it makes no sense to seek the optimal solution; instead it's more effective to look for a satisfactory one. In my own research, I have extended his work on psychological AI to a wide range of situations of uncertainty.

Fast-and-Frugal Trees

When making decisions, experts tend to use fewer pieces of information than novices; they know what is relevant and ignore the rest. And if some cues (features) are more important than others, experts consider these first and may base their decision on the most important cue only. My research team and I have programmed these intuitions into simple algorithms called *fast-and-frugal trees*, named for their speed and frugal use of information.

In conflict-ridden countries, innocent civilians are frequently wounded or killed after being mistaken for terrorists at military checkpoints. The main problem is that checkpoint personnel, who have to infer whether an approaching vehicle contains civilians or suicide attackers, typically get no systematic training in how to make these life-or-death decisions. Together with armed forces instructors, some of my research colleagues designed a fast-and-frugal tree that helps make more reliable decisions.[15] The first question in the tree is whether an approaching vehicle contains more than one occupant (figure 2.1, left side). If yes, the inference is that the occupants are nonhostile civilians (because assigning multiple suicide attackers to one car would be a waste of scarce resources). If not, the next question is whether the vehicle slows down or stops at the checkpoint. If not, the inference is that the occupant is hostile. If yes, the third and final question is whether further threat cues exist (such as an intelligence report about a suspicious green Honda Civic in the area). The tree is easy to memorize and execute, and it can reduce civilian casualties by more than 60 percent.

Another risky venture is to identify failing banks. Traditional finance bets on highly complex "value-at-risk" models that purport to safely estimate the capital a bank needs to avoid a heavy loss with a high probability, say 99.9 percent. Yet these models did not prevent forty-two of 116 large global banks (with more than $100 billion assets at the end of 2006) from failing during the 2007 financial crisis. Part of the problem is that these models are too complex and fragile for the uncertain world of banking—they need to estimate millions of risk factors and their correlations on the basis of often unreliable data. Together with experts at the Bank of England, my colleagues and I developed a fast-and-frugal tree that matches or outperforms complex methods in predicting bank failure (figure 2.1, right side).[16] The first branch of the tree asks for the leverage ratio (roughly, the ratio of a bank's capital to its total assets) of each bank and is placed first because this

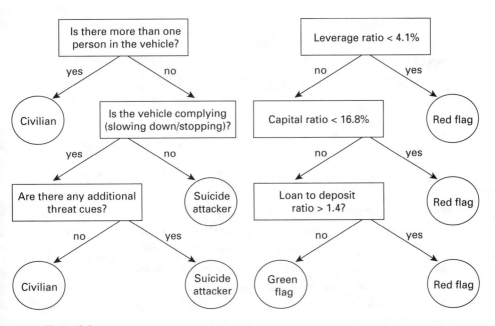

Figure 2.1
Psychological AI. Fast-and-frugal trees are examples of psychological AI, which constructs decision aids based on psychological principles. Left: A fast-and-frugal tree for checkpoint decisions that reduce civilian casualties. Right: A fast-and-frugal tree for identifying banks that are in financial distress. In general, a fast-and-frugal tree has a small number of cues (or questions) and allows for a decision after each cue. From Katsikopoulos et al., *Classification in the Wild*.

ratio discriminated best between the banks that failed and those that survived. For instance, the Swiss bank UBS, which had to be bailed out by the Swiss authorities during the crisis, had a leverage ratio of only 1.7 percent and would have immediately been given a red flag by the simple algorithm. UBS would have satisfied the other two features in the tree, but the logic of fast-and-frugal trees is that each question stands alone in the order of its importance, and one cannot compensate a bad value with good values on the other cues. This is akin to the functioning of human bodies and other complex systems: perfect kidneys cannot compensate for a failing heart.

Psychological AI, such as fast-and-frugal trees, can augment and improve human decisions. In each example, expert knowledge is transformed into an algorithm. Unlike many more complex algorithms, psychological AI is transparent, which allows its users to understand and adapt an algorithm when situations change. In situations of uncertainty, both human judgment and transparency are indispensable.[17] In the case of banks, transparent algorithms that make no room for estimating millions of risk factors can help authorities more readily detect when banks are trying to game the rules.

Gaming AI

The stable-world principle helps to understand what problems an AI application is likely to be more successful at solving. We need to know more than that, however, in order to evaluate whether it is likely to be successful in the real world. Much of AI is commercial or military, not scientific. Commercial organizations may have goals that conflict with what their product is officially supposed to do. Even in a stable world, where an AI product could be delivered to the benefit of society, the AI may be gamed to serve a hidden interest. The problem is not the technology but rather who is behind it. I illustrate three ways customers are taken in, both in situations where AI works and where it does not: how a potentially useful AI application is exploited for profit, how a mediocre algorithm is made to look impressive, and how ineffective technological solutions are marketed even when investing in people would be more effective.

How Electronic Health Records Were Gamed

A medical record contains all relevant information about a patient's medical history, such as test results, diagnoses, and treatment results. These confidential files have been traditionally kept on paper. One problem with paper is that when a patient visits a new doctor, it is difficult to access the patient's history in the time available, and many tests are unnecessarily repeated. That raises costs and takes away doctors' valuable time from listening and talking to the patient. Moreover, if a new doctor is not aware of relevant aspects of a patient's medical history, the patient may be given unintentionally harmful treatment. To avoid this, electronic health records, also called electronic medical records or electronic patient records, promise to be fertile tools for fast access to the information that doctors need. These records have algorithms for recording and storing information, including images, which the doctor can access quickly, unless the file is hundreds of pages long. Keeping records and making these readily accessible is an ideal task for basic AI programs that involves sorting past data in a (hopefully) reliable way.

At a summit meeting in 2003, British Prime Minister Tony Blair boasted to US President George W. Bush about his new multibillion investment "Connecting for Health" to wire the entire British health care system. Back in Washington, Bush pressed for a similar program to computerize heath records.[18] Researchers at the RAND Corporation estimated that the US health care system could save $81 billion per year by adopting it rapidly.[19] That estimate led to the US surpassing Great Britain's investment, with a $30 billion federal incentive program and an explosive growth of the industry. Finally, we might think, a case where investment in a database paid for patients. However, when the RAND Corporation reviewed the situation in 2013, their optimism faded. Hospitals with electronic record systems had *increased* their billing. Costs had gone up instead of down, from about $2 trillion aggregate expenditures in 2005 to $2.8 trillion in 2013.[20] Moreover, RAND reported that the quality and efficiency of health care were only marginally better, if at all.

Why did such a good idea not pan out? The answer is that the system was gamed.

Consider the hope of saving costs. The plan was that easy access to past tests would reduce unnecessary duplicate testing. In fact, when a doctor entered data into the record, the software automatically recommended new procedures to consider. After being prompted, just to be on the safe side, doctors using electronic records ended up performing more, not fewer, tests.[21] Almost all of these alerts appear to have been false alarms.[22] Companies were gaming the software for sales pitches to increase profit and inflate medical bills.

Consider the hope of better access. The plan was that electronic records would allow doctors and patients to access all needed health information quickly and at any time. The generous funding, however, led to competition between companies, and the systems installed by hospitals and doctors were proprietary, employing incompatible formats and secret algorithms that were not designed to communicate with another.[23] Thus, access was in fact limited, not universal—just as you can't use the charging cable of a Mac for a PC or even for your previous-generation Mac. The competing software companies had used the government subsidies to increase their own gains. Their primary goal was brand loyalty, not patient safety.

Consider the hope of improving patients' health. As mentioned, there is lack of evidence that patients benefit from their electronic health records.[24] Some software systems even prompt doctors to "upcode" a diagnosis to more severe ones, which increases billing and profit and leads to more tests and treatments. All this happens at the expense of increased anxiety among patients, who are oblivious to the fact that their severe diagnosis may have been made in the interest of the billing system. Not all doctors are aware of these issues. One reason is that the vendors of electronic health reports can protect themselves from liability by nondisclosure agreements in their contracts, meaning that doctors and clinics who experience flaws in the software are prohibited from discussing these openly.[25] Doctors who have spoken up on safety issues, serious injuries, and deaths related to problems with the software later requested that their names not be disclosed for fear of being sued.[26]

The sad fact is that electronic health records were developed by software companies for maximizing billing, not patient care. Hospitals and doctors have used the government subsidy provided by Medicare HITECH to buy electronic systems that allow them to charge Medicare even more for their services. Discussions on electronic health records, however, rarely recognize this issue, and instead mostly revolve around issues of privacy and doctors' time. Privacy is indeed a legitimate concern when hundreds of thousands of doctors, health administrators, and insurance personnel have access to patients' personal files. Clinics have been hacked, records have been stolen, surgeries have had to be canceled, and clinics have been blackmailed to pay for getting their records back. Recently, even patients have become the target of hackers: tens of thousands of patients from psychotherapy centers in Finland were blackmailed with the threat of having their intimate conversations with their therapists published online.[27]

These and the fact that managing electronic health records decreases doctors' time with patients are serious problems.[28] Yet solving them requires first solving the more fundamental problem that the potential benefit of electronic patient records for the patient is eaten up by a primarily profit-driven system. There are too many players—clinics, administrators, lobbyists, big pharma, insurers—who pursue interests conflicting with those of patients and their health. Such interests do not provide a fertile ground for using AI to serve patients. To reap the fruits of digital health, we need health systems that first and foremost serve patients.[29] Otherwise, digital health will become a Band-Aid solution or will even amplify the problems.

The Texas Sharpshooter Fallacy

Electronic health records could be beneficial for patients; the data are known and only need to be made accessible. Now consider a situation of uncertainty rather than relative stability: predicting divorce. The stable-world principle indicates that algorithms are unlikely to perform well here. Yet there is a clever way to make these predictions look much better than they are. This procedure is sometimes a fallacy, that is, an unintentional result out of ignorance, but also can be a deliberate trick to persuade people.

Predicting Divorce

Assume you are newly married. Would you like to know whether your marriage ends in divorce? One might object that this is a hypothetical question, because whether a couple will split is hard to know ahead. Yet in a series of studies, clinical researchers reported they had found an algorithm that predicts with about 90 percent accuracy whether a couple will divorce in the following three years.[30] Study after study claimed the algorithm had high levels of accuracy, varying between 67 and 95 percent.[31] These impressive findings attracted wide attention in the media, and love labs and marriage institutes worldwide advertised "science-based predictions" of relationship stability and therapy.[32] Unlike in online dating agencies, the algorithm in these clinical studies is transparent. It uses features such as education, the number of young children, violence, binge drinking, and drug abuse.

Who would have thought that an algorithm could tell whether a young couple will divorce? In the future, you can drive up to the love lab with your partner, get interviewed, and then be told whether the two of you will likely get divorced or not. If the answer is yes, you might as well contact an attorney before wasting any more time together. Or, better yet, consult the love lab before even tying the knot. But think for a moment. When it comes to marriage, we are not dealing with a stable world. Whether a couple stays together depends on so many factors, loaded with plenty of uncertainty. That alone is reason for skepticism about the algorithm's level of accuracy. There is indeed a delusion behind these numbers. And it can be found in many a study in the social sciences, not only in predicting divorce. Let me explain.

The Sharpshooter

A Texas cowboy shoots at a barn wall with a revolver from quite a distance. He proudly shows his amazing performance: the bullet holes are grouped around the bull's-eye of the target. How did he do that? Years of hard training? A miraculous revolver?

None of that; the strategy is much easier. What the sharpshooter in fact did was shoot first and then paint the circles of the target around the bullet holes so that the bull's-eye was in the middle (figure 2.2). Clearly this procedure guarantees better results than if the cowboy had painted the target first before shooting. If one counts all shots that "hit" the target, then fitting the target to the shots produces nine out of ten possible hits, or a 90 percent accuracy. Doing it the right way—that is, by painting the target first, say at the center of the barn—would have likely led to fewer hits. You might think of this trick as cheating, and it is. In science it is called *data fitting*, which by itself is nothing immoral. But to use the eye-catching results and sell them as *prediction* is clearly deceptive.

In our analogy, the sharpshooter is the algorithm, the target is the prediction the algorithm makes, and the bullet holes are the data. To shift the target into the optimal position, the sharpshooter has two degrees of freedom: move it left or right and up or down on the barn wall. With these two options, he can center any possible set of shots miraculously around the bull's-eye. These degrees of freedom are called the *free parameters* of an

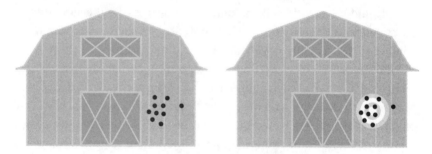

Figure 2.2
Shoot first, paint target later. The Texas sharpshooter shoots first (left) and then draws the target around the bullet holes so that the bull's-eye is in the center (right). The sharpshooter's accuracy looks impressive if one doesn't know that the bullet holes (the data) were obtained first and then the target was fitted to the holes.

algorithm. Algorithms can have more than two free parameters. Financial models, for instance, have considerably more, which gives them huge flexibility to account for almost anything—after the fact.

If the sharpshooter had first painted the target on the barn and then neatly shot around the bull's-eye, he would indeed deserve admiration. Along these lines, real prediction occurs when predictions are made first, and then the data are obtained. The distinction between fitting and predicting is the lifeblood of every student in machine learning and should be evident to scientists in every field. But in the social sciences or in commercial enterprises, algorithms are often merely fitted to match the data, and the result is sold as prediction.[33] Here is a useful principle:

Always check whether an algorithm's impressive performance was obtained by prediction or mere fitting.

Let's go back and apply this principle to the notable levels of accuracy in "predicting" divorce. In fact, none of these studies had actually predicted anything at all. The authors always knew which couples were still together and which had divorced, and then, like the Texas sharpshooter, fitted their algorithm to the data.[34] What they should have done instead is a two-step procedure: take one group of couples and fit (develop) the algorithm, and then test the predictions for a new group of couples.

Independent researchers did precisely this with a new group of 528 people.[35] When they fitted the algorithm to the data, it "predicted" a similarly high percentage (65 percent) of the divorced couples correctly. When the algorithm was tested on a new group of couples and *did not already know* whether any of the marriages had ended in divorce, everything changed. The algorithm correctly predicted only 21 percent of those who would get divorced.[36] That is its true accuracy.

And there is another question: How good is 21 percent? Is it better than a blind guess? Among the couples studied, 16 percent got divorced in the first three to six years of marriage. That means if you close your eyes and simply predict that each couple will get divorced, you can already expect to get 16 percent right. That is the benchmark. Any algorithm worth its salt must therefore predict substantially better than that. In the study we've been looking at, the algorithm does a bit better, but not much. The true performance of the algorithm was 5 percentage points above chance; all the rest was hot air.

Beware of the Texas sharpshooter fallacy. Even when the term *prediction* is used, as in the studies on the divorce algorithm, it may be falsely used. To find out, check whether the algorithm has been first trained on one sample of data and then tested on a different sample. This two-step process is called *cross-validation*. Many neuroscientists, psychologists, sociologists, economists, and educational researchers just fit their algorithms to data, then stop and report the big numbers.[37] You might wonder why. One reason is that this produces more impressive numbers. Reporting these electrifies audiences unaware of the Texas sharpshooter fallacy and persuades them of the magical powers of an algorithm. There is also a more charitable but even more alarming explanation. Quite a few social scientists do not appear to understand that fitting is not prediction, a confusion that has a long history and has been documented in many other studies.[38] At the beginning of the twenty-first century, merely fitting data was the rule in psychology.[39] Today, social scientists increasingly begin to recognize the importance of making true predictions.

Financial charlatanism is another instance of the sharpshooter fallacy. You may have heard great news like this: "Our backtesting shows: This innovative strategy beat the market over the last ten years on average by 5 percent." Backtests are historical simulations of an investment strategy. As in the case of the divorce algorithm, all the data are known, and large numbers of investment strategies are fitted to the data; the one with the highest fit is then advertised as the best investment strategy—similar to the Texas sharpshooter's strategy. This problem has been exposed many times, but many in the world of finance proceed by fitting their algorithms to data, without always reporting how many algorithms they tried.[40] Being honest would not produce attention-grabbing numbers, and customers might prefer to invest in the competition's algorithms. The solution is not to install a code of honesty, which would put the honest ones out of business. Instead, it would be to have informed citizens who ask whether a Texas sharpshooter strategy was involved.

The physicist Niels Bohr liked to say, "Prediction is hard, especially about the future." This remark has been attributed to Mark Twain, Yogi Berra, and a host of others known for their wit.[41] But it is in fact quite serious. As we have seen, what is called prediction may have nothing to do with the future. Hindsight is easy, but prediction is difficult, particularly in matters of love and divorce.

The Moon Shot

You may have seen IBM's commercials where what looks like a sentient box interacts with Bob Dylan, Serena Williams, and other celebrities. Watson's claim to fame began in 2011, after winning the TV quiz show *Jeopardy!* against two of its best contestants. After this spectacular success, the value of IBM's stocks increased by $18 million.[42] Ginni Rometty, IBM's CEO, announced that the "next moon shot" would be health care—not because Watson knew anything about health but because that is where the big money is. To adapt Watson to health care, it was fed medical data, such as patients' case histories and treatments. Before it was clear whether Watson could actually learn to diagnose and recommend treatments like a doctor, high expectations were raised. Rometty announced a medical "golden age" in which AI "is real, it's mainstream, it's here, and it can change almost everything about health care."[43] IBM's PR department created the impression that Watson would revolutionize health care, or had already, and the Watson team came under pressure to quickly commercialize their product.[44] The first application was oncology.

Watson for Oncology became marketed around the world for recommending treatments for cancer patients. From the MD Anderson Center in Texas, one of the most respected cancer centers in the United States, to Manipal Hospitals in India, clinics purchased the services and paid a per-patient fee ranging from $200 to $1,000. Yet Watson could not even deliver the performance of a capable human doctor, let alone the moon shot. Many of the program's treatment recommendations proved to be incorrect and unsafe, endangering patients' lives.[45] IBM's messages eventually faded from hype to modesty, announcing that Watson's medical knowledge was at the level of a first-year medical student.[46] MD Anderson annulled its contract, realizing that the software could not live up to the claims of IBM's marketing department. The cancer center had spent $62 million, which made Watson the best-paid med student ever. Major German clinics also fired Watson after realizing that its treatment recommendations bore more resemblance with artificial stupidity than intelligence. In the words of the CEO of Rhön-Klinikum, a cooperation of hospitals and clinics, to carry on with Watson would be the equivalent of "investing in a Las Vegas show."[47] Other IBM partners have also halted or shrunk Watson's oncology-related projects. IBM has yet to publish any scientific papers demonstrating how well the technology affects physicians and patients.

The moral of this story is not about Watson or AI in general. It is about aggressive marketing that cannot fulfill the expectations raised and about the uncritical faith in the marketing hype by journalists, unlike among AI scientists. In the words of Oren Etzioni, CEO of the Allen Institute for AI and former computer science professor, "IBM Watson is the Donald Trump of the AI industry—outlandish claims that aren't backed by credible data."[48] In more polite terms, Watson is just a computer program that could morph into an assistant for routine medical tasks, but not the brilliant doctor advertised.

IBM also markets Watson as a general-purpose intelligence in espionage, law, and finance. Naïve bankers buy Watson's services to make better investment decisions. But if Watson is such a crack at predicting the stock market or simply making smart investments, IBM should not be in the financial difficulty it is.

Make People Smarter

Why not invest more money and keep trying? That is a legitimate question— except that too much trial and error eats up the attention and resources that could better be spent on effective measures that truly save lives from cancer. Robert Weinberg, a world-renowned cancer biologist at MIT, has devoted his professional life to seeking causes and cures for cancer. Yet cancer drugs may prolong life by only a few weeks or months while massively decreasing its quality. Moreover, they are so expensive that no country in the world can afford them for all of their citizens. At a 2011 conference in Amsterdam hosted by the Royal Dutch Academy of Sciences and the Dutch Central Bank, Weinberg gave a keynote with a surprising message for a cancer biologist. Despite having spent his life unraveling the biological conditions of cancer, he views the real promise in fighting cancer today outside of biological science: in making children and teenagers health literate. The argument is this:

• About half of all cancers have their roots in behavior. Smoking, diet, and lack of movement leading to obesity are the key factors.

• These habits are set early in childhood or in adolescence.

- Thus, we need to start early in making children and teenagers health literate to prevent these habits that cause cancer to be formed.

Weinberg and I joined forces and designed an experimental program in schools, where young people are not told what to do and what not to do but instead learn what the health risks are and how they will be lured by advertising and their peers into unhealthy behavior. The program teaches skills such as the joy of cooking, knowledge of how one's body functions, healthy activities, basic scientific attitudes such as asking questions and finding out answers by doing experiments, and awareness of where to look up trustworthy information. Rejecting compulsion or nudging, it provides young people with tools for taking control of their health. We featured this risk literacy program in two subsequent conferences on preventing cancer in Amsterdam. The head of the Dutch Cancer Society visited me in Berlin and told me he was interested in funding the program in schools in areas of the Netherlands where childhood obesity is on the rise.

At the third conference on reducing cancer incidence, the very same head of the Cancer Society gave a talk. We had expected to hear him speak on the program for preventing cancer. Instead he spoke about the promises of big data for curing cancer. Weinberg and I could not believe our ears. Afterward, we talked to him, but to no avail. The head of the society had been persuaded otherwise, not wanting to fall behind all the other organizations that fund big data research. That was the end of our project to make young people health literate. All the funding went to the industry.

3   Machines Influence How We Think about Intelligence

The most profound technologies are those that disappear. They weave themselves into the fabric of everyday life until they are indistinguishable from it.
—Mark Weiser, "The Computer for the Twenty-First Century"

I believe that at the end of the century [i.e., by 2000] the use of words and general educated opinion will have altered so much that one will be able to speak of machines thinking without expecting to be contradicted.
—Alan Turing, "Computing Machinery and Intelligence"

After Deep Blue and AlphaGo became kings of the game board, many felt that humans were no longer the crown of evolution. A flood of popular writers began to talk of a coming superintelligence, a singularity, and the melting of brains with computers into immortal cybercreatures. The most pessimistically inclined announced the end of humanity, with us becoming pets on the leash of robots, exhibits in zoos, or the first species to use its intelligence to extinguish itself. All these stories make the same mistake. Both enthusiasts and pessimists equate computing power with human intelligence.

How did the view arise that intelligence is what computers do? One might expect there to be solid proof, but it's based on analogy, not evidence.[1] New technologies have always inspired new psychological theories. There is nothing wrong with that as long as one recognizes the analogy for what it is and doesn't confuse it with the real thing. Similarly, the human tendency to anthropomorphize may make many especially keen to attribute human intelligence to nonhuman entities.

Consider the marvels of human memory. People can learn to recite tens of thousands of lines of the Mahabharata of ancient India or conduct entire symphonies by heart. To understand how this is possible, a wide array of analogies has been used, from wax tablets to libraries. The most influential of these were new technologies, which suggested that memory is like a hard disc, a telephone switchboard, or a hologram. The same holds for our understanding of the marvels of human intelligence. While we know how a hologram or a computer works, we have only a limited understanding of how intelligence functions. So we take what we know and try to understand what we don't. And there are indeed similarities: both the computer and the human nervous system are electrical; artificial neural networks also have networks of connections, and brains have a neural network of synapses. But every analogy has its limits. If the mind were a computer, we could calculate the square root of 1,984 in a fraction of a second. If a computer were a mind, it could just as easily pass the CAPTCHA for proving that you are not an algorithm. Moreover, it could not only win at chess but also have fun while playing.

Although minds are not computers, the creation of this analogy is a fascinating story. Computers themselves have been inspired by human systems; in fact, the computer was originally modeled after a new social system, the division of labor.

The First Computers Were Social Systems

The French Revolution set out to destroy a society of aristocrats who lived glamorous lives, spent millions on rich banquets, and could not have cared less if the heavily taxed peasants perished from hunger. One side effect of the revolution was the attempt to make our measurement systems more rational: to introduce the decimal system for measuring weight, height, and almost everything else (up to dividing days into ten hours, with a hundred decimal minutes, each of which had a hundred decimal seconds, but that was going too far). The new system required calculating logarithmic and trigonometric tables, difficult calculations that had previously been the task of mathematical prodigies. Yet the French Revolution also brought about a new vision of calculation. Inspired by Adam Smith's *The Wealth of Nations*, the engineer Gaspard de Prony invented a social hierarchy with a three-tiered division of labor. At the top were a few famous mathematicians who devised the formulas, in the middle seven or eight persons trained in analysis, and at the bottom seventy to eighty unskilled persons who simply added and multiplied millions of numbers.[2] The world had never seen such a project before.

Some years later, the English mathematician Charles Babbage (1791–1871) was amazed that elaborate calculations could be performed by an assembly of workers knowing so little. Going from there, Babbage conceived of replacing the workers with machinery and set out to build his first digital computers. These did not really work, but eventually over the following century functioning calculators were built. My point here is that computers and concepts of intelligence are deeply interwoven. In the beginning there was a new social organization of work, and the computer was created in its image.[3] The first computers were humans.

How Calculation Became Divorced from Intelligence

A computing system with unskilled workers and Babbage's mechanical machines clashed with deeply held beliefs. In the seventeenth and eighteenth centuries, during the Enlightenment, the ability for mental calculation was considered the sign of a brilliant intellect. In the psychology of the time, the innovative mind constantly took apart ideas and rearranged them into new ones. Thought was understood as a combinatorial calculus, and great thinkers were proficient calculators. The story of the mathematician Carl Friedrich Gauss (1777–1855; figure 3.1) is probably the best known. Gauss came from a poor family in Brunswick, Germany, and was nine years old when his primary schoolteacher posed an addition problem to the class:

> Schoolmaster Büttner, a whip in hand, told the pupils to write down all numbers from 1 to 100 and add them up. Within a short time, the young Gauss wrote down the answer on his slate, and said, in the town's dialect: "There it lies!" Disparagingly, but with some compassion, the teacher looked at the little scamp, who apparently had the cheek to claim he had found the result to a problem that cost the entire class bitter sweat. Young Gauss explained to the teacher how he had reasoned. The hundred numbers make 50 pairs (1 + 100, 2 + 99, and so on), each of which adds up to 101. Therefore, the sum is 50 times 101, which is 5050. To the teacher's credit, he realized Gauss's genius and concluded "the boy can learn nothing new in my school." Büttner ensured that Gauss was sent to a better school.[4]

Prodigious mental reckoning was a favorite story in the eulogies of great mathematicians. But the story of Gauss's brilliance in arithmetic was one of the last. The demonstration that calculation could be done by unskilled

Figure 3.1
The German ten-mark bill featured Carl Friedrich Gauss and included the formula for the normal distribution he devised. The bill was issued in 1991 and circulated until the euro was introduced in 2002. The normal distribution is the basis of many models in machine learning, economics, psychology, and beyond. Gauss was one of the last mathematicians who were celebrated for their excellent mental reckoning. After it was shown that complicated calculations could be done by a social system of unskilled workers and even by machines, calculation was no longer considered the hallmark of a genius.

workers and by machines slowly put an end to its glorification. No longer was it considered the hallmark of a genius. When calculation shifted from the company of great men to that of the unskilled workforce, the psychology shifted as well. What was once considered the crown of human intelligence sunk to a mechanical level. What became known as intelligence in the nineteenth century had little to do with calculation anymore. This change set the stage for the new romanticism of the twentieth century that declared creativity to be a mystical process and sneered at calculation. Even today, high-level politicians and celebrities can count on getting sympathy votes by boasting about their poor grades in math. Sir Alec Issigonis, famous for having designed the Mini car, once commented, "All creative people hate mathematics. It is the most uncreative subject you can study."[5]

Women Computers

Throughout history (and sometimes even today) occupations that lost their prestige were opened to women. After calculation became looked down upon as a mechanical, mindless task, it was delegated to women. Up to the

Figure 3.2.
"Women computers" processing data from the sixteen-foot wind tunnel in NASA's Ames Research Center in 1943. © NASA.

second half of the twentieth century, social systems continued to perform large-scale computation. The Manhattan Project at Los Alamos, where the atomic bomb was built in the 1940s, relied on "human computers," despite having all the technology for calculation available. These were mostly low-paid women, who repetitively performed the same calculations such as cubing a number and passing it on to another person, who incorporated it into another computation.[6] The fact that women did the calculations reflected a shift in the social status of calculation at that time.

Women ultimately staffed the calculation departments in major statistical projects, from observatories to insurance offices to military research (figure 3.2). When calculating machines became more reliable and even indispensable, instead of the machines being viewed as more intelligent, the human calculators were viewed as more mechanical.[7] Calculating machines added up numbers, just as washing machines cleaned laundry and sewing machines made stitches—all machines that were typically operated by women. The thought that these machines embodied artificial intelligence occurred to no one.

The low prestige of calculation also explains why computer science became a field open for women, as opposed to the rest of engineering. In the early 1980s, almost 40 percent of US graduates in computer science were female. Yet in that decade, when personal computers became available and their ownership prestigious, they were marketed to men who enjoyed playing games, and parents were likely to buy them for boys, not girls. Since then, the presence of women in computer science departments has drastically declined, down to 18 percent by 2020.[8] Another factor that may have driven women out is that computer science evolved from insignificance to a highly paid profession. Today, the heads of tech companies are immensely rich and almost all male.

Intelligence Becomes Computation

The demotion of calculation from an intellectual to a mechanical activity took a 180-degree turn when electronic computers became available in research centers. In his last work—the 1956 Silliman lectures, which he could neither finish nor deliver due to his premature death by cancer—John von Neumann, one of the greatest mathematicians of his time, pointed out similarities between the nervous system and the computer, that is, between neurons and vacuum tubes, but added cautionary notes emphasizing their differences. While von Neumann observed parallels at the level of the hardware, Alan Turing called the observation that both cells and diodes are electrical a "very superficial similarity."[9] In his view, the important similarities lie in the function, not the hardware. Turing asked whether machines can think; his famous "imitation game," popularly known as the Turing test, evaluates whether a machine can imitate a human. Going even further, he considered whether computers could have free will and contemplated teaching machines by using the same principles used to teach children. Note that Turing asked whether the computer is like a mind, while psychologists, led by Herbert Simon, later asked the opposite question: whether minds are like computers.

In the 1960s and 1970s, Simon, together with computer scientist and psychologist Alan Newell, argued that human intelligence is akin to a computer program. In their view, known as the *physical symbol hypothesis*, intelligence consists of the operation of symbols over time.[10] The symbols are manipulated by rules, generating new symbols as outputs over time. For instance, logic is a system containing symbols such as *and* and *or*, and its rules are those of logical deduction. Similarly, chess is a system where the symbols are the pieces, and the rules are the legal chess moves. In a digital computer, the symbols are zeros and ones that define memory, and the rules are those of the central processing unit (CPU) that change memory. According to Simon, all these symbol systems resemble the human mind and brain.[11] This view inspired many a movie featuring robots disguised as humans that think just like us, only faster and with less emotion, and that, with sufficient computing power, could eventually turn into superhumans.

Simon famously told a graduate class in early 1956 that, over Christmas, he and Newell had invented a thinking machine and "thereby solved the mind-body problem, explaining how a system composed of matter can have the properties of mind." As with the first large-scale social computing systems in the French Revolution, the prototype of their first computer program, the Logic Theorist, was made of human components: Simon's wife, children, and graduate students.[12] That same year, the term *artificial intelligence* went public at the now-famous Dartmouth AI Conference. It was coined by the computer scientist John McCarthy, a staunch supporter of free speech who initiated *time-sharing systems*, later to be renamed *servers*, now called *cloud computing.*

Importantly, Simon realized that the computer is more than a calculator; it can manipulate any symbol, from text processing to videos. In hindsight, it is clear how revolutionary the symbol-processing idea was and hard to understand why this potential was not obvious earlier. For instance, in 1952, Howard Aiken, a pioneer in computing who designed Harvard's Mark I calculator, reminisced, "Originally one thought that if there were a half dozen large computers in this country, hidden away in research laboratories, this would take care of all requirements we had throughout the country."[13] We may smile at that anecdote, yet the original idea behind the computer—and the internet later on—was to solve problems in science and engineering, and little more.

How was Simon ever able to convince others that cognition is computation, given that the related idea that intelligence is calculation had been discredited for over a hundred years? When Simon first proposed the idea, no one paid any attention. It was too early. Computers at that stage were large-frame machines that proved difficult to use; getting a program to work could take months. For instance, in an average week in the years

1965 and 1966, the PDP computer at Harvard's Center for Cognitive Studies saw eighty-three hours of use, fifty-six of which were spent on debugging and maintenance. Rather than a marvel of intelligence, the computer was a steady source of frustration, as witnessed by the desperate title of one of the center's technical reports from 1966: "Programmanship, or how to be one-up on a computer without actually ripping up his wires."[14]

But in the 1970s and 1980s, electronic computers became fast and cheap, and began to populate the desks of psychologists and other researchers. Psychologists finally accepted the view of the mind as a computer only after computers became indispensable tools in their daily routine. With these new desktop tools, the old Enlightenment link between intelligence and calculation—as illustrated by the story of Gauss—was resurrected in a new form. The nature of intelligence came to be seen as computation, that is, what computers do. Ever since, psychologists have spoken about cognition as computation. What else could it be?

The idea that cognition is computation was based not on new evidence but on new tools and inspired by analogy.[15] Thus, computers inspired new theories of human intelligence, just as centuries ago the new social organization of work made mechanical computers thinkable. With today's heavy-duty computing power, this idea morphed into the AI-beats-humans argument, the vision of an artificial superintelligence that soon will match human intelligence in all respects while surpassing the human brain in speed and efficiency. What we see is a new version of the old Enlightenment view that intelligence is calculation.

At the same time, breakthroughs with deep neural networks have revived the belief that AI could encompass all of human intelligence and create a superintelligence capable of doing everything humans can and more. In the next chapters, we will take a closer look at the "intelligence" of neural networks and see how fundamentally it differs from human intelligence, so that such a fear (or hope) remains science fiction. Let us begin with a practical, mundane dream of automatization.

4 Are Self-Driving Cars Just Down the Road?

I am extremely confident that Level-5 [self-driving cars] or essentially complete autonomy will happen, and I think it will happen very quickly . . . I remain confident that we will have the basic functionality for Level-5 autonomy complete this year.
—Elon Musk, Tesla CEO, July 2020

Autonomous vehicles are nowhere near as smart as they need to be. Safety features—including manual override—must be top priority.
—US Senator Richard Blumenthal

On a dark desert night in March 2018, Elaine Herzberg was pushing her bicycle across a four-lane road in Tempe, Arizona. Herzberg was forty-nine years old and homeless, carrying bags on her bicycle. She had already crossed three lanes when she was struck and killed by an Uber autonomous car.

What should we think about the promise that self-driving cars will soon steer us safely through regular traffic? For one, Uber's cars were not even self-driving; they had human drivers at the wheel as backups.[1] The Volvo that struck Herzberg in Arizona was a test car with a backup driver and had been in autonomous mode for nineteen minutes before the accident. Traveling at forty-three miles per hour, the car's radar detected Herzberg six seconds before impact. Yet the AI's "perception module"—which builds a model of the environment using cameras and radar—got confused and classified her first as an unknown object, then as a vehicle, and finally as a bicycle. Each of these led to a different predicted path, and for more than four seconds the system did not make the inference that braking was needed. Only one second before impact did it do so—and was unable to come to a halt because the braking technology had been disabled by the Uber engineers.

The reason was previous bad experience, where autonomous cars braked for no good reason and were then rear-ended by human drivers. It was instead the driver's task to hit the brakes. But instead of watching the road, she was apparently looking down at her smartphone and streaming *The Voice*, a popular singing contest.[2] Video footage shows her looking up with a shocked expression immediately before the collision. Who is responsible when a purported self-driving car kills a person? The chief of the Tempe police blamed the homeless victim. Uber avoided a lawsuit by a secret out-of-court settlement with the victim's family within days of her death and later was cleared of criminal wrongdoing. The backup driver faces manslaughter charges.

Herzberg had been part of a technology experiment without knowing it. Before her tragic end, California regulators had made it difficult for Uber to test their cars in real traffic. Doug Ducey, governor of Arizona, seized the opportunity and tweeted: "Here in AZ we WELCOME this kind of technology & innovation! #ditchcalifornia #AZmeansBIZ."[3] The fact that the cars and drivers were unsafe was known. Before striking Herzberg, an autonomous car was involved in a collision nearly every other day on the road. Five days before the accident, a manager in Uber's autonomous vehicle unit had sent an email to the company's executives warning that the AI in their prototype robotaxis was dangerous and that backup drivers were not properly trained.[4] Nevertheless, the state of Arizona and Uber carried on with the experiment. After the deadly accident, Arizona's governor quickly changed his mind and banned Uber's experiments. This time, the governor of Ohio grabbed the chance to lure a big tech company away from Silicon Valley and welcomed it to "the wild, wild West" of unregulated self-driving-car testing.[5] The upshot of this sad story is a race to the bottom of regulation and safety.

The Adapt-to-AI Principle

A self-driving car is one that can drive *everywhere* and *under all traffic conditions* without a human as a backup. The hope is that electronic chauffeurs will dramatically reduce traffic accidents while increasing convenience. Car manufacturers such as Elon Musk (see epigraph) suggest that this prospect is just around the corner.[6] Yet traffic conditions involve considerable degrees of uncertainty: the erratic behavior of human drivers who do not always follow rules; cyclists, jaywalkers, or animals appearing on the road; driving at night when the headlights of other cars are blinding; and weather conditions such as heavy rain and ice. Given these uncertainties, the stable-world principle questions the widespread commercial fairy tales that self-driving cars are just down the road.

The confusion between what technology can actually deliver and what people are made to believe begins with a careless use of the terms *self-driving*, *driverless*, and *autonomous cars*. This imprecision has fueled unrealistic hopes. Let's clarify the terms. The Society of Automotive Engineers' classification system distinguishes between five levels of automation (figure 4.1). Level 1, long in existence, includes adaptive cruise control and lane-keeping systems as driver assistance tools. A Mercedes van was one of the first vehicles to demonstrate these skills on public roads, on the German autobahn in 1986.[7] Level 2 combines various Level 1 technologies to automate more complex driving tasks, such as automatic parking, and is the state of the art of commercially available cars. It requires drivers to keep their hands near the wheel and *constantly* pay attention in case something unexpected happens. A Level 3 system is a big technological jump and can perform most driving tasks but still needs a human in the driver's seat who is receptive to alerts from the system and ready to take control if something goes wrong. At all three levels, humans have to be constantly alert in case something unexpected happens.

The next two levels operate without a vigilant human at the wheel. At Level 4, cars are able to drive fully autonomously without human help but only in restricted areas, such as motorways, airports, and factories, or in specifically designed cities, but not everywhere and in all conditions. Level 5 refers to self-driving cars, that is, cars able to drive safely under the full range of driving conditions without any human backup, including all weather, road, and traffic conditions.[8] These five levels are often not distinguished, creating the illusion that fully self-driving cars are already zipping along the roads.

Self-Driving Cars Require Stable Environments

The uncertainty involved in driving a car can be understood by comparing it with flying a plane. Aviation introduced automation long ago: today,

SAE J3016™ LEVELS OF DRIVING AUTOMATION

	SAE LEVEL 0	SAE LEVEL 1	SAE LEVEL 2	SAE LEVEL 3	SAE LEVEL 4	SAE LEVEL 5
What does the human in the driver's seat have to do?	You <u>are</u> driving whenever these driver support features are engaged – even if your feet are off the pedals and you are not steering			You <u>are not</u> driving when these automated driving features are engaged – even if you are seated in "the driver's seat"		
	You must constantly supervise these support features; you must steer, brake or accelerate as needed to maintain safety			When the feature requests, you must drive	These automated driving features will not require you to take over driving	
	These are driver support features			These are automated driving features		
What do these features do?	These features are limited to providing warnings and momentary assistance	These features provide steering OR brake/ acceleration support to the driver	These features provide steering AND brake/ acceleration support to the driver	These features can drive the vehicle under limited conditions and will not operate unless all required conditions are met		This feature can drive the vehicle under all conditions
Example Features	• automatic emergency braking • blind spot warning • lane departure warning	• lane centering OR • adaptive cruise control	• lane centering AND • adaptive cruise control at the same time	• traffic jam chauffeur	• local driverless taxi • pedals/ steering wheel may or may not be installed	• same as level 4, but feature can drive everywhere in all conditions

Figure 4.1

Five levels of automation in cars, as defined by the Society of Automotive Engineers (SAE). The self-driving car is at Level 5. It is defined as a car that requires no human backup and is able to drive safely everywhere and in all traffic conditions. Current commercial cars operate at Level 2. The stable-world principle suggests that safe driving at Level 5 will be extremely difficult to achieve, while Level 4 is likely by restricting autonomous cars to specific areas and adapting our streets and cities to AI. Source: © SAE International from SAE J3016™ Taxonomy and Definitions for Terms Related to Driving Automation Systems for On-Road Motor Vehicles (revd. June 15, 2018), https://www.sae.org/standards/content/j3016_201806/.

algorithms are what fly planes most of the time. In fact, about half of the costs of a new commercial plane are spent on software validation and debugging. However, the software of a large commercial airplane is less complex than that needed for a self-driving car.[9] An autopilot system may have to deal with one or a few other aircrafts in its vicinity and has time to act because of the large distances between them. The software for a self-driving car needs to deal with dozens of cars, cyclists, and pedestrians, and their proximity means that critical decisions about life and death need to be made within fractions of a second.

The stable-world principle has an important implication, the *adapt-to-AI principle*:

> *To improve the performance of AI, one needs to make the physical environment more stable and people's behavior more predictable.*

If we hand over decisions to algorithms, we have to adapt our environments and our behavior. This may involve making humans transparent to algorithms, regulating human behavior, or even eliminating humans from the playing field. Applied to autonomous cars, the two principles give us a clue about what is likely to happen:

> *There will be no self-driving cars (Level 5 automation). Rather, a fundamental change will happen: our cities and roads will be redesigned to create the stable and predictable environment that algorithms need (Level 4 automation), such as wired highways from which human drivers are banned, and cities where human driving is illegal.*

In other words, if we want autonomous cars, we have to adapt to their potential. The technology is not just a support system; it requires *us* to adjust our behavior. Such behavioral shifts have happened before.

When cars were invented by Carl Benz in 1886 and mass-produced by the Ford Motor Company in the early twentieth century, the transition from the horse age to the motorized age proved to be dangerous. Cars had to share traffic space with horse-drawn vehicles, bicyclists, pedestrians, and children playing on the road, without any regulations. For cars to function safely, paved roads and strict traffic rules were needed. Human behavior in turn became restricted by drivers' licenses, license plates, traffic lights and signs, and rules such as driving on a particular side of the road. That was not the first time. Already in 1831, the Paris police had issued an ordinance that traffic—carriages, horses, mules, and pedestrians—should keep to the right to reduce the urban chaos.[10] This rule was followed as grumblingly as was an earlier ordinance that people should stop emptying their chamber pots out the window onto the street. Clean streets allow for free-flowing traffic. The modern highway, with its smooth surface made of asphalt or concrete, is the ideal environment for cars. Pedestrians, bicyclists, and animals are banned. New technologies don't simply assist us; their effect is deeper. To benefit from their maximum efficiency, we have to be prepared to adapt—to change our behavior and our environments.

Neural Networks in the Driver's Seat

How might engineers build an electronic chauffeur to whom one would entrust one's life? As explained in the previous chapter, there are two visions. Psychological AI would teach computers to drive the way humans drive, that is, replicate the human process of perception, judgment, and decision-making. This is the spirit of good old-fashioned AI (often abbreviated as GOFAI): to study how humans solve a task, then translate human behavior into "production rules" and program these rules into a computer. An example of such a rule is "if a child steps onto the street in front of the car, brake immediately." Yet this idea is doomed in regular traffic because it would bog the AI down with too many rules—including those that tell it how to distinguish a running child from a floating plastic bag. Memorizing thousands of rules is also not the way humans learn to drive. In fact, what we do when seamlessly driving is impossible to describe precisely enough to set down as rules for an algorithm. Research therefore turned to building machine-learning systems. This second vision of AI does not attempt to teach machines to drive like humans but uses algorithms to best solve a specific problem, such as recognizing traffic signs and moving objects. The ways machines solve these problems have little or nothing to do with human problem-solving and are best known from search engines and recommendation systems.

The big breakthrough in automation—and the big hope for truly self-driving cars—came with a type of algorithm called *deep artificial neural networks*. These are fundamentally different from the algorithms for finding true love or creditworthiness. Computer systems that drive cars are made of three modules. The first, a perception module, consists of cameras, radar, and lidar, which is a version of radar that uses invisible pulses of light. Cameras spot traffic signs, traffic lights, lane markings, and other features. Radar measures the velocity of objects, and lidar the shapes of the surroundings. This information is used to build a model of the car's environment. Object recognition constitutes a major part of the progress made toward autonomous cars. Second, a prediction module forecasts what the "recognized" objects are going to do next, such as in what direction nearby cars and pedestrians will move, and how fast. Third, a driving policy module relies on these predictions to decide whether the car should speed up, slow down, or veer to the left or right. In the case of Herzberg's death, the policy module did not decide to slow down because the perception and prediction modules were confused.

Artificial neural networks are sometimes presented as magical machines that only a computer whiz can comprehend. But there is a nontechnical way to get a general idea of how they function. Similar to algorithms used for partner matching and credit scoring, they consist of three parts: an input, its transformation, and an output. Consider a network that is being trained to recognize a school bus. Pictures of objects form the input, only some of which are school buses. The pictures are digital, that is, an image is divided into thousands or millions of pixels, which is exactly what a digital camera does. Every pixel is analyzed in terms of three color values: red, green, and blue. From the perspective of an artificial network, an image is nothing more than a huge table of numbers, each specifying one of the three colors for a pixel. This table is called the *input layer*. The *output layer* consists, in the simplest case, of two numbers: 1 for school bus and 0 for all other objects. The big difference to classical AI lies in the way the input is transformed into the output.

Unlike in a love algorithm, the input is not transformed into an output by a (secret) formula. Rather, the input layer is transformed through a series of hidden layers, which make the network deeper—hence the name *deep neural network*. Such a network may have just a few or hundreds of hidden layers. A layer deeper down in the network can contribute to refining the previous layers. This fine-tuning is also called *deep learning*.

Deep neural networks can "learn" in three ways: supervised, unsupervised, and through reinforcement. Supervised learning is the best known, so called because there is a "teacher" who provides yes/no feedback but no rules. It is used for recognizing objects such as a school bus and traffic signs. To learn properly, the network is given thousands or even millions of photos as well as their correct classifications, and then learns a function that matches the input with the output as correctly as possible. In recommendation systems, such as when Amazon suggests books to buy, you are the teacher who clicks or likes certain contents while browsing. Unsupervised learning takes place without feedback. For instance, the network searches for similarities in pictures and generates clusters of similar pictures. This form of learning corresponds to gazing at the stars at night in quest of patterns without knowing a thing about astronomy. And the third form, reinforcement learning, occurs in tasks such as Go where the network gets feedback not after each move but only at the end of the game. In this case, all the network knows are the rules of the game and the goal, and it has to find its own way to get there. AlphaZero's amazing performance is due to reinforcement learning, yet it may also result in finding unexpected shortcuts. One algorithm that was developed to play Tetris simply learned to pause the game to avoid losing.[11]

Although it is relatively easy to understand the general logic of deep neural networks, it's much more difficult to understand what a network exactly "sees." Herein lies a major difference between it and psychological AI, where the rule is transparent—at least to the designers if the algorithm is secret. What transpires in the hidden layers of a deep neural network, however, is extremely hard to figure out, even for the engineers who built the network. You may have heard the mortifying story that Google's image classification system identified a dark-skinned couple as "gorillas." The engineers immediately reacted, but not in the way you might think. Instead of attacking the heart of the problem, the silent solution was to remove the categories *gorilla*, *chimpanzee*, and *ape*. Since then, pictures of apes are no longer classified as such.[12] Deep neural networks have grown too complex to be easily understood and corrected.

Traditional computer vision, in contrast, relies on individual features such as edges and colors. For objects such as a school bus, it remains a challenge to write down a rule that defines it exactly. Rules work better for recognizing handwritten numbers, where features may include lower round strokes, as in 5 and 6, and upper round strokes, as in 2 and 3. In the case of deep neural networks, it is hard to tell whether the individual units in the various layers extract individual features or work together in collective actions.[13] What we do know is that the view that artificial neural networks function akin to traditional computer vision or human vision, only faster, has been proven wrong. We will see this below in the form of the astonishing errors that neural networks make that are alien to human intuition.

Deep neural networks are nothing new. Artificial networks have been known since the 1950s and 1960s. In fact, some statisticians have complained that engineers with little knowledge of statistics have reinvented statistical tools already in existence and clothed these in a misleadingly impressive new jargon, such as *artificial intelligence*.[14] Their complaints are not fully justified; several technical innovations did spark the revival of neural networks. These include the advance in the number-crunching power of digital computers, the development of efficient training methods, the large sets of data now available, and the extraordinary power of graphics processing units that were originally developed for video gaming. Still, the fact that most of supervised and unsupervised learning has long been known helps make it clear that there is ultimately no different quality of "intelligence" in deep neural networks than in ordinary statistics.[15]

Despite the suggestive term *neural network*, networks do not "perceive" and "think" like humans. Their similarity to the neural circuits in the brain is actually quite shallow. Algorithms can solve tasks that are beyond human capacity, but they also have bizarre blind spots and can make mind-boggling errors. The following example brings home what these blind spots can mean for self-driving cars.

Do Neural Networks Know What a School Bus Is?

Autonomous cars need software that can reliably identify what's on the road: traffic signs, as well as people and vehicles. On the left-hand side of figure 4.2 is a picture of a typical yellow school bus in the United States that a deep neural network has learned to identify correctly. The middle picture shows a multicolored pattern of pixels. In reality, the pattern is so tiny that it has been magnified here by a factor of ten for the reader to see it. The picture to the right of it was generated by superimposing this pattern upon the first picture. To the human driver, the school bus on the right-hand side is still a school bus, and special attention is required when sighted because kids might be around. For the deep neural network, in contrast, the school bus has disappeared. It now classifies the object as an ostrich.[16]

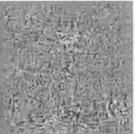

Figure 4.2
Deceiving neural networks. Left: A picture of a school bus that is correctly recognized
by a deep neural network. Middle: A pattern of pixels (magnified here by a factor of
ten to make it visible). Right: The pattern in the middle is added to the picture of the
school bus on the left, resulting in the picture to the right. The neural network no
longer recognizes the school bus but classifies it as an ostrich (Szegedy et al., "Intriguing Properties of Neural Networks"). The color version can be accessed at https://
arxiv.org/pdf/1312.6199.pdf.

What constitutes a minute continuous change for humans, almost invisible, such as the difference between the two school buses, is *qualitatively*
different for the machine.[17] This kind of error is alien to human intuition.
Such qualitative "leaps" are not limited to image recognition: they also
have been discovered in voice, speech, and text recognition.[18]

The reverse phenomenon also exists, where a deep neural network cannot see a difference between two pictures that are strikingly different to
human eyes. Consider the picture of the school bus in figure 4.3 on the
left. It is the same bus as in figure 4.2 (left) that a trained deep neural network correctly recognized. To the right of the bus is an image of horizontal orange-and-black stripes, whose colors roughly resemble those of the
school bus. After the network learned to identify school buses correctly, it
was tested on the picture with the stripes. The network classified the picture
as showing a school bus. It was 99 percent certain of being correct.[19]

These striking errors are not limited to the specific kind of deep neural
network tested. Nor are they limited to school buses; they crop up for virtually every object tested. One might think that new technology can eliminate
these kinds of errors, but that's not the case, and for two reasons. First, new
technology may be able to fight individual errors but not their causes. Second, new technology is constantly being developed to find additional ways
of fooling deep neural networks. An arms race between protective measures

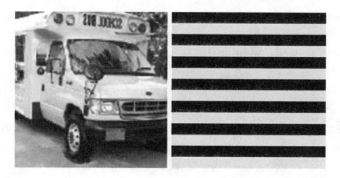

Figure 4.3
A deep neural network mistakes stripes for a school bus. The picture on the left (same as in figure 4.2, left) is correctly classified as a school bus. But the network is also 99 percent certain that the picture to the right is another school bus. The stripes and the school bus are similarly colored. Based on Szegedy et al., "Intriguing Properties of Neural Networks," and Nguyen et al., "Deep Neural Networks Are Easily Fooled." © IEEE.

and adversarial algorithms has evolved, including successful one-pixel attacks, that is, low-cost deception by modifying a single pixel in an image.[20]

Why do these errors occur? A first thought is that the neural networks had too small a sample of pictures of school buses to learn from, or only untypical samples in order to deliberately trick them. But neither is the case: the networks were trained on standard image datasets comprising more than one million images; adding more training photos would not have helped. The answer lies in the fundamental difference between human and machine intelligence. Humans have a mental concept of a bus. Children can recognize a bus after seeing just one or a few. A bus has four wheels, headlights, and a windshield, is larger than a car, and is designed to carry many people. A deep neural network has no such concept of a bus, nor does it know its function. It does not even know where the bus is in the image.[21] Its intelligence consists of finding statistical associations between pixels and assigning probabilities to pixels. The stronger the association, the more certain the network. If typical elements of objects are repeated in a picture, the network will therefore likely grow more confident that the object exists in the picture. The striking yellow of a school bus framed by black parts appear to be the typical elements it has detected. These colors are amplified by the horizontal stripes.

Similarly, a deep neural network has no concept of a zebra but searches for typical patterns in images, such as stripes. As a consequence, a picture of a zebra with two additional legs (and thus more stripes) would startle a human but may make a deep neural network more confident that the zebra is indeed a zebra.[22] Why not simply teach a network that zebras cannot have more than four legs? Fixing this specific error will not eliminate the underlying problem: *The network doesn't know that a picture represents something in the real world; it has no concept of things.* Its intelligence is confined to detecting patterns in colors, texture, and other features. Human intelligence, in contrast, is about representing the world.

A Color Patch on a T-Shirt

To make a self-driving car safe, a deep neural network needs to recognize objects such as vehicles, pedestrians, and traffic signs correctly (the *perception module*), but it also needs to predict where cars or pedestrians are moving next (the *prediction module*). Networks that make these predictions are called *optical flow networks*. Typically, cameras behind the windscreen take many pictures per second and infer from these the location of the vehicles in the next time frame. Predicting where other vehicles or bicycles are heading needs to be impervious to little disturbances; otherwise safety cannot be guaranteed.

However, as in object recognition, deep neural networks can make inferential errors that are totally nonintuitive for humans. A small color patch, like the example in figure 4.4, can baffle the network when it estimates the flow of traffic.[23] Even a patch that covers less than 1 percent of the entire picture can completely erase the motion the network "sees" in half or all of the picture. The larger the patch, the more disturbing its effects. Such patches can be placed on pedestrians' T-shirts, a car bumper, or a traffic sign. It takes only a few hours to generate these patterns. If the type of network is known, then it is easier to generate the patterns, although the one shown in figure 4.4 works for a variety of networks, that is, even if the kind of network a self-driving car uses is unknown.

No human would be confused by such a pattern when driving. The errors humans make are of a completely different nature. An estimated 90 percent of crashes are due to human factors such as driving well above the speed limit, using handheld electronic devices, ignoring fatigue, and driving while

Figure 4.4
An example of the color patches that confuse neural networks' predictions of the flow of traffic when placed on pedestrians' T-shirts or on the rear of a car. These patches are "universal" in the sense that they confuse not a particular network but major classes of networks. It is sufficient if a patch covers less than 1 percent of the entire picture. © The Max Planck Institute for Intelligent Systems 2019. For the color version of this and other adversarial patches, see https://arxiv.org/pdf/1910.10053.pdf.

drunk.[24] No network would fail because it is tired or distracted by an incoming phone call—although running out of power could have a similar effect. The conspicuous differences in errors here and for object recognition show how machine intelligence differs fundamentally from human intelligence. It is often said that a computer knows only what it has been told. That may be true for psychological AI but not for networks. A deep neural network can learn to "know" things humans have a hard time understanding.

What does all of that mean? Engineers who build vision systems can be fooled by assuming that when the AI is trained on thousands of pictures of, say, traffic signs and then performs well on a test set, it will also recognize these in new situations in the same way a human would. The blind spots can lead to accidents that are difficult to foresee and comprehend. Moreover, malicious characters can trick deep neural networks into making mistakes. For instance, a car's computer vision can be trained to mistake a stop sign for a 65 mph speed limit sign, and thus send car and passengers hurtling through an intersection at top speed.[25] Similarly, pedestrians who make unexpected fidgety movements such as dancing on the street can confuse deep neural networks.[26]

Intuition and Morals

Three-year-old children already understand that people, unlike objects, have intentions and desires and routinely infer intentions from other people's gaze, bodily movement, or tone of voice. The ability to attribute intentions to others is also known as *theory of mind*. It is used to foster safe driving. Consider a child standing on the curb of an urban road. Human drivers can infer in a blink whether the child intends to run out onto the street: if the child's gaze is on a ball on the other side of the street, that could well happen; if the child's gaze is instead directed at a woman close by, it is unlikely. Humans have learned to infer intentions using such heuristic rules. Artificial neural networks have none of these intuitions. Even if the AI could reliably detect a child and an adult woman with its perception module, it would additionally need an intuitive psychology module that can infer their intentions and their likely next actions. How to program intuitive psychology into a machine remains a riddle. That would require a true breakthrough in software engineering.

If self-driving cars did exist, a huge set of moral dilemmas would arise. Consider three elderly people crossing a road on a red light (figure 4.5). A self-driving car recognizes them too late to brake in time and is about to kill all of them. The car's only other option is to steer into a wall and kill its three occupants instead. How should the engineer program the driving policy module of the AI? Whom should the car kill?

In 2017, the German Ethics Commission on Automated and Connected Driving was the first worldwide to propose ethical rules for this problem.[27] These include that human lives should have priority over animals' lives and that discrimination by age, gender, or any other personal feature

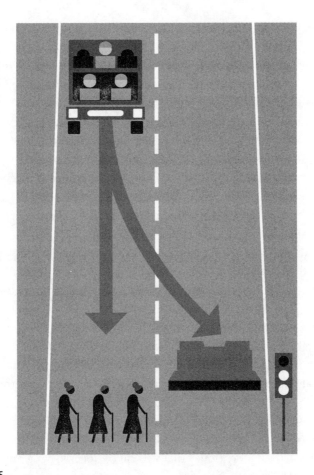

Figure 4.5
What moral intuitions should an autonomous car be given? In this hypothetical
scenario, a self-driving car cannot avoid an accident and must choose who should
live and who should die. The choice is between killing three elderly pedestrians who
are crossing the street on a red light or saving the pedestrians' lives by crashing into
a wall and killing the car's three occupants (see Awad et al., "The Moral Machine
Experiment").

should be prohibited. By implication, the car should be indifferent to saving pedestrians or saving passengers. A study with millions of people in over 200 countries found that most disagreed with these ethical rules, apart from sacrificing dogs and cats instead of humans. For the scenario with the three elderly pedestrians, the far majority voted that the AI should be programmed to kill the pedestrians rather than the car occupants. After all, they are old and are disobeying the law by jaywalking. In general, the majority of people worldwide showed rampant discrimination and did not consider their fellow humans equally.[28] People would generally save the life of a human above a dog's but not if the human was a criminal. Similarly, people of lower social status and homeless people were considered less worthy to be spared.

Moral dilemmas, even if hypothetical as in this study, can provide a window into the values of a culture. People in Western countries had the strongest preference for inaction, that is, avoiding the responsibility of actively killing people such as by turning left and crashing into the wall. People in Eastern countries such as Japan and China and in Islamic countries such as Saudia Arabia and Indonesia showed the strongest preference for saving the elderly or pedestrians and for sacrificing the unlawful. In contrast, people in Central and South American countries most strongly preferred saving females, the young, and individuals of high social status.

The moral dilemma of the self-driving car is a version of the "trolley problem" in moral philosophy. This class of thought problems is named after a fictional runaway trolley that will kill five people lying on the tracks unless one takes action, such as by pushing a fat man over the bridge and onto the track so that he stops the trolley, killing him in order to save five lives. These problems can reveal whether someone prefers inaction and letting many lives be killed or actively takes responsibility for the death of a single person to save other lives. But what is most telling is that these moral experiments assume a world of certainty; they shy away from the uncertainty of real life. Their assumption is that the fat man will not defend himself and will fall precisely at the right spot and at the right second on the track. Similarly, even though the AI in the hypothetical scenario has no time to avoid the accident, the assumption remains that the AI has perfect recognition skills, will properly categorize the individuals' ages and behavior, and will be able to determine precisely how many of the pedestrians or car occupants will die. When this hypothetical certainty is replaced by realistic uncertainty, people's firm moral judgments become more permissive and less stringent.[29]

Moral psychology has a long tradition of posing moral dilemmas that ignore uncertainty. As a thought experiment, that is not a problem. But in real traffic, certainty remains an illusion; both human and artificial intelligence are forced to make decisions without knowing exactly what the consequences will be. This key difference may spare engineers from having to worry about how to program a machine to deal with this kind of moral dilemma, and in fact, most engineers have discarded the illusory certainty of the trolley problem as a guide to ethics.[30] The hypothetical dilemma of choosing whom to kill may never become reality because it assumes fully self-driving cars with perfect prediction of what will happen after a collision.

The Future of Driving

As mentioned before, contrary to Elon Musk's suggestive claims in this chapter's epigraph and to other commercial promises, the stable-word principle suggests that self-driving cars are unlikely to be our future. Safe Level 5 cars would require true breakthroughs in software development. Automatic driving will come, but not in the way we have been led to believe. Instead, it will likely come in two forms, one of which allows humans to drive (Level 3), the other not (Level 4). These two forms are likely to coexist, and in both instances, humans need to adapt to the potentials of the AI.

When Your Car Reports You to the Police

If self-driving cars are not going to happen, one alternative appears to be training humans to use AI as a support system but to stay alert and retain control if it fails—which is called *augmented intelligence*. It amounts to partial automation, that is, to sophisticated versions of Level 2 or 3. Yet augmented intelligence entails more than just adding useful features to your car and may well lead us into a different future, where AI is used to both support and surveil us. That possible future is driven more by insurance companies and police than by car manufacturers. Its seeds are in telematics.

Young drivers are reckless, overconfident, and an insurance risk, according to the stereotype. Some indeed are, but many are not. Nevertheless,

Feature (Event)	Weight
Rapid acceleration or harsh braking	40
Driving over the speed limit (more than 20% for more than 30 seconds)	30
Driving at night between 11 p.m. and 5 a.m.	20
Driving in cities	10

Figure 4.6
Pay as you drive. A black box in the car reports four features of the driver's behavior to the telematics insurer. Safe driving is calculated by an algorithm whose features and weights are transparent. The insurer offers better rates if the customer drives safely.

insurers often treat them as one group and charge a high premium. Telematics insurance can change this by offering better rates for safe drivers. The idea is to calculate the premium from a person's actual driving behavior instead of from that of the average driver. To do so, a black box that connects to the insurer is installed in the car (using a smartphone is possible and cheaper but less reliable). The black box records the driver's behavior and calculates a safety score. Figure 4.6 shows the scoring system of one of the first telematics insurers. It observes four features and assigns them different weights.[31]

Rapid acceleration or harsh braking is assigned the greatest weight, followed by driving over the speed limit. Each driver starts with a monthly budget of 100 points for each of the four features. An "event" results in points being subtracted, such as 20 points for the first rapid acceleration or for driving over the speed limit. At the end of the month, the remaining points are weighted as shown and summed up to a total safety score. Although telematics is often called black box insurance, the algorithm is not at all a black box like most love algorithms. It is explained in detail on the insurer's website, and everyone can understand and verify the resulting score.

Personalized tariffs are advertised as promoting fairness. They do so by taking individual driving style into account. But they also create new sources of discrimination when driving at night and in cities is punished. Hospital staff, for instance, may have little choice to avoid working at night and in cities. Thus, some of the features are under the driver's control, but not all. Interestingly, one feature that is under the driver's control is absent in virtually all personalized tariffs: texting while driving.

And the black box that allows fairness also enables surveillance. Consider a possible future. Why should the black box send a record of speeding only to the insurer? A copy to the police would be extremely handy and save them much effort. It would make all speed traps obsolete. If you speed, the car prints out the ticket on time or, more conveniently, deducts the fine automatically from your online account. Your relationship to your beloved car may change. There is a slippery slope between fairness and total surveillance.

Would you be in favor of a new generation of cars that send traffic violations directly to the police? In a survey I conducted, one-third of the adults said yes, more so among those over sixty and less among those younger than thirty.[32] The technology for this future already exists, as most new cars come with a black box installed. The data it collects do not belong to the car owner and can be used in court against the driver. In Georgia, the police obtained black box data without a warrant after a deadly accident, and the driver was found guilty of reckless driving and speeding.[33]

While the motives for surveillance vary, digital technology supports all of them. One need not even buy telematics insurance. Modern cars have built-in internet connections, and—without it being made transparent in the owner's manual—most send their car manufacturer all the data they can collect every couple of minutes, including where the driver currently is, whether harsh braking occurred, how often the position of the driver seat was changed, which gas or battery-charging stations were visited, and how many CDs and DVDs were inserted.[34] Moreover, as soon as you plug in your smartphone, the car may copy your personal information, including contacts' addresses, emails, text messages, and even photos. Car manufacturers are remarkably silent about this activity, and when asked with whom they share this data, they typically do not reply.[35] That information helps to find out many other things of interest, such as how often drivers visited McDonald's, how healthily they live, and whom they occasionally visit at night. Connected cars can support justice and improve safety but also spy on you. Telematics insurance embodies the double face of digital technology: surveillance in exchange for convenience.

Cities Adapt to Autonomous Cars and Driving May Become Illegal
In the possible future where cars surveil their drivers and report traffic violations to the police, drivers will adapt their behavior. Future Level 2 and

3 drivers will be predictable and law-abiding. Speeding, running red lights, texting while driving, drunk driving, not wearing seatbelts, and other traffic violations will become things of the past. In this future, humans are still in the driver's seat.

As mentioned before, the adapt-to-AI principle lets us expect a second possible future that may unfold in parallel. It requires rebuilding the current traffic environment to a more predictable one that allows for Level 4 driving—cars without humans at the wheel but not in all traffic conditions. To begin with, special lanes on highways and areas within a city may be restricted to autonomous cars only. In the long term, adapting to the potential of self-driving cars may require building entirely new cities that provide cars with a stable environment. That may entail outfitting roads with sensors that detect cars and systems that communicate with cars. Human drivers may be banned from the roads, and pedestrians may be given their own secluded pathways. In this future, human driving may become illegal in restricted areas, or entire cities, and humans may eventually lose the ability to drive. Living in this future, you would look back in disbelief to the past when people were allowed to drive themselves and kill others.

One vision is to integrate autonomous cars with public transport. A fleet of robotaxis would circle through a city at low speed for safety reasons. These vehicles would transport people to bus and streetcar stations, where they could continue traveling faster and for longer distances. That would make privately owned cars obsolete, and cities packed with parked cars history. The space could be used for wide lanes for pedestrians and bicyclists, physically separated by walls or hedges to prevent them from entering car lanes. To encourage this development, use of robotaxis and public transport could be made cheap, private car ownership extremely expensive, and private cars banned from entering cities.

Some countries have already taken steps toward this future and are building new cities around Level 4 technology. For instance, Toyota announced its plan to build a smart city near Tokyo and Mount Fuji that is adapted to autonomous cars and where pedestrians will be consigned to walkways out of reach of the cars.[36] The autonomous cars will drive citizens into the center of town, Toyota says—although that may also be an obsolete idea, given that there won't be much to do downtown anymore. Streaming replaces movie theaters, online shopping replaces shopping expeditions, and autonomous cars deliver medication, pizza, and everything else you may desire to your doorstep. Similarly, the Chinese government announced its intentions to build a city the size of Chicago south of Beijing that can support autonomous traffic. Such radical changes to the public sphere are presently unthinkable in the United States and Western Europe, but these countries may also soon adapt their concepts of civil rights to reap the benefits of AI by providing it with an environment in which it can flourish.

These two possible futures of autonomous driving illustrate a more general point. The question is how to make AI work in situations of uncertainty. One solution is 24-7 surveillance and behavior modification through immediate reward and punishment. This method makes humans more predictable. AI can more easily deal with people who follow rules and behave consistently, whether on the road or elsewhere. The second solution is to make the environment more stable and predictable. That means adapting it to the potential of the AI, which may require removing pedestrians, cyclists, and human drivers entirely from the streets.

Bike-First Cities and Smart Public Transport

A radical alternative to these two visions of autonomous driving is to do away with cars, autonomous and otherwise, as far as possible. One prospect is bike-first cities. Amsterdam, Copenhagen, and other cities have already shaken up their infrastructure and redesigned their streets around bikes, not cars. The city of Utrecht in the Netherlands transformed itself from a traditional car-friendly city to one that gives bikes the upper hand over motorized traffic.[37] Ninety-eight percent of households own bikes. Car lanes have been converted into broad and safe bike roads, and traffic lights are adapted to the speed of cyclists so that they can glide through the city without constant stops. As a result, the city has become much less noisy, the air has become cleaner, and fewer pedestrians and cyclists are killed in traffic. The city invests some $50 million every year into bike infrastructure and saves an estimated $300 million every year thanks to reduced pollution and health care costs. "You really have the idea that people are the boss of the city, not machines," commented the city's vice-mayor.

Bike-first cities combine well with the second alternative to smart cars, smart public transport. Long-distance bullet trains such as the Shinkansen between Tokyo and Osaka, the Fuxing Hao from Shanghai to Hangzhou, France's TGV, Italy's AGV, and the German InterCity Express operate at 150 to 250 miles per hour. The Shinkansen, for instance, is silent, safe, very comfortable, and punctual to the second. The world's fastest short-distance train hits over 260 miles per hour and takes only seven minutes from Shanghai's Pudong International Airport to a metro station in Shanghai. From there, a tight inner-city transport network above and below ground gets you near where you want to be, and the rest can be done by walking or taxi. Electric trains are faster and safer than cars and are less harmful to the environment.

The original focus on private personal cars rather than public transport is probably not an accident. The self-driving car industry is centered in California and other states where well-connected networks of public transport barely exist today after being decimated by car manufacturers in the 1930s and 1940s, unlike in Hong Kong, Seoul, Singapore, London, Paris, Berlin, Madrid, or New York. Not only does public transport barely exist, but it is often looked down upon as a system for the poor and underprivileged groups. Fifty-six percent of Americans reported that they never use public transport, compared with 2 percent of Russians, 2 percent of Chinese, 4 percent of South Koreans, 12 percent of the British, and 15 percent of Germans.[38] This cultural bias contributes to the enthusiasm about cars and may create a blind spot for alternative solutions.

The Russian Tank Fallacy

Much of AI is about prediction, as in the perception and prediction modules of autonomous cars. In chapter 2, I introduced the Texas sharpshooter fallacy, where an algorithm is fitted to match the data, and its performance is then reported as prediction, without any actual prediction having been made. This makes the algorithm look much better than it is. While this fallacy exists in the social sciences, it is virtually absent in machine learning. When training neural networks to recognize images, it is easy to avoid the sharpshooter fallacy by using cross-validation: dividing the set of pictures randomly into two subsets, and then training the network on one subset and testing it on the other.

Deep learning, however, can run into a second problem, which stems from the fact that we do not know what a neural network has learned. As seen above, a network has no concept of a school bus or a right-of-way sign. Instead, it learns cues that discriminate between classes of objects in a set of pictures, even when these may be irrelevant and do not generalize to other pictures or the real world. The picture of the horizontal yellow-and-black stripes illustrates how a network can miss the mark. The problem is this: because the network has no understanding of concepts as humans do, it may focus on irrelevant cues that are present in *both* the training and test set.

I will refer to this second problem as the *Russian tank fallacy*. The name stems from an urban legend that goes like this:

> The US Army trained a neural network to tell Russian tanks from US tanks. As usual, a large data set with photos of tanks was used; the network was trained on one half and tested on the other. Finally, it could master the test with 100 percent accuracy. But when it was used in the real world, it flunked. Only later was it realized that the U.S. tanks had been photographed on a sunny day and the Russian ones on a cloudy day. The smart network had found the perfect cue: clouds.[39]

Similar stories have been told about neural networks that learned to tell wolves from dogs. In the dataset of pictures, the dogs happened to be standing on grass and the wolves on snow. The network found the best differentiating cue: white background. Although these stories may contain more legend than truth, the general point is that although networks are excellent at detecting cues on how pictures labeled as "Russian tank" or "wolf" differ, these cues may have little to do with humans' concept of a tank or a dog.

In other words, a researcher in machine learning who avoids the sharpshooter fallacy can still fall into a new trap, ignoring the fact that the network may learn perfect but irrelevant cues. Many studies in machine learning do not make the next move and test the AI's performance in the real world. In health care, for instance, most algorithms are tested on computers, not in hospitals. The assumption that what works in the lab will work in the real world has been called "the big lie in machine learning."[40] For instance, one proposed solution has been convolutional neural networks (CNN) that have a built-in bias, such as edges and other structures indispensable to the object in question. But as the following real case illustrates, such networks do not always solve the problem.[41]

At New York's Mount Sinai Hospital, one of the oldest teaching hospitals in the US, a deep neural net (CNN) learned to make accurate predictions from chest X-rays as to whether patients were at high risk of pneumonia. At other hospitals, however, it flopped. Eventually, the cause was found by manually inspecting the X-ray images. The AI had learned to detect whether these were taken by the hospital's portable chest X-ray machine or in the radiology department. Portable X-rays were taken of patients too sick to leave their rooms, who had a greater risk of lung infection. In other hospitals, that cue was of little help. Detecting irrelevant cues is one of the reasons why a neural net trained in one hospital typically becomes less accurate if used in another hospital.[42]

The Russian tank fallacy relates to a much deeper issue, the question of external validity. Many machine-learning studies proceed by saying, "Look, here are some real-world data, let's split them into two parts and see how accurately AI predicts the part it has not yet seen." In some quarters, this has become the only method used. While it takes care of the sharpshooter fallacy, it is not the same as predicting the future. In fact, by splitting the data randomly in two, one creates a fairly stable world: the data used to teach the algorithm and the data used for the test are from the same group. Nothing unforeseen can happen, and no future events are predicted. For instance, some lauded studies about computer-based predictions of personality based on Facebook likes and other digital footprints do not actually make any true predictions about the future.[43] In general, machine-learning systems that predict well in one setting but not in others have been observed in many instances, such as with patients in different hospitals, images taken by different cameras, and biological assays from different cell types.[44] To overcome this limitation, one more step is needed: to test how well an algorithm performs in new situations, such as with new patients in different hospitals. That is true prediction.

The general lesson is: software that displays spectacular results in a computer lab can lose its potency when tested in the real world. If you hear that an algorithm successfully identified children crossing the street or patients at risk of lung disease, check whether the algorithm was tested in the real world. Dividing pictures into two sets, training a network on one set, and testing it on the other is important, but not sufficient. What counts is testing it in the wild.

5 Common Sense and AI

The intuitive mind is a sacred gift and the rational mind is a faithful servant. We have created a society that honors the servant and has forgotten the gift.
—Albert Einstein

Common sense in an uncommon degree is what the world calls wisdom.
—Samuel Taylor Coleridge, *Literary Remains*

A human brain is a miracle, with its nearly 100 billion neurons and 100 trillion connections, fine-tuned over hundreds of millions of years of evolution. These are numbers that our conscious mind can scarcely comprehend. Human brains are also highly energy-efficient compared with contemporary computer technology. Consider energy, which is measured in units called watts in honor of James Watt, the British engineer and inventor of the steam engine. The supercomputer Watson used 85,000 watts when playing *Jeopardy!*, compared with the 20 watts used by each of its two human competitors' brains—as much as a dim light bulb.[1] The brain's main energy source is glucose. Watson also needed tons of air-conditioning equipment, while its competitors could have done with a handheld fan on the hottest days of the year. One of the world's most powerful computers, the Blue Waters supercomputer at the University of Illinois, consumes about 15 million watts and occupies 20,000 square feet of floor space. It requires a large cooling system hidden underneath the floor and is not mobile. The brain needs little space: it is the size of two fists and easy to carry around. And unlike the supercomputer, it is self-assembled and able to program itself. If all human brains were replaced by supercomputers, the amount of heat generated might well put a swift end to our climate.

The Cogs of Common Sense

As mentioned before, human intelligence evolved to deal with uncertainty. Among the mental skills humans have evolved to succeed in an uncertain world, four stand out:

Causal thinking. The ability to think in terms of causes develops early. Children persistently ask "why?" They want to know why the sky is blue, why some people are rich, and why they have to eat their vegetables. Some children pose more why questions than their parents can answer. By doing so, children build causal models of the world. Being curious about causes rather than mere associations is characteristic of human intelligence and is the hallmark of science. Causal thinking is both a strength and a source of superstition, as when believing that crossing your fingers will bring you luck.

Intuitive psychology. Children develop an intuitive psychology in the first years of their lives: they "know" that other humans have feelings and intentions, and they can take the perspective of another human.[2] Special brain circuits appear to be dedicated to monitoring what other minds know, think, and believe. Lack of intuitive psychology is a sign of autism.

Intuitive physics. Similarly, children develop an intuitive physics and understand the basics of time and space. For instance, they know that a solid object cannot move through another object, that objects persist over time, and that time cannot be reversed.

Intuitive sociality. When children are older than three years of age, they are motivated to follow group norms (cooperation and competition) and develop and defend moral standards.

I refer to the sum of these skills as *common sense*:

Common sense is shared knowledge about people and the physical world enabled by the biological brain, and requires only limited experience.

Common sense arises from a mixture of genetic predispositions and individual and social learning (such as knowing that the world is three-dimensional, or that one should not hurt others' feelings). It can be exercised intuitively or by deliberate judgment. For instance, most people can reliably tell a genuine smile from a merely polite one but cannot explain how they do it. That is called intuition, or a gut feeling.[3] However, when one learns that in a genuine smile, the muscles around both the mouth and the eyebrows move, while in a polite smile only those around the mouth move, that insight enables a conscious judgment. Intuition and judgment are not opposite poles. They rely on the same process: here, the same visual cues.[4]

Common sense is a huge challenge to those in the business of developing artificial intelligence. That holds even for the basic understanding that words and pictures represent objects in the social and physical world. We have not succeeded in programming common sense into computers with the help of rules or by creating deep neural networks that are able to learn it. Sensory motor skills are another big challenge. Although AI can beat a human at chess, the same program cannot take the chessboard down from the shelf and set up the figures. It is still difficult to build a robot that can move its fingers as flexibly as a violinist can, or a housekeeper robot that can carry out all household chores as well as a human can. In the absence of these skills, one solution is to redesign our living spaces to fit the abilities of AI.

Computers excel in a different set of abilities:

Speedy calculation. Fast calculation is the lifeblood of search engines and chess computers.

Finding associations in big data. Increasing speed of calculation also enables the search for associations among large sets of variables.

Detecting patterns in images or acoustic information. Algorithms can detect patterns in pictures, such as in genomes and astronomical observations, which are difficult for the human eye to spot.

Fast computing power by itself, however, produces neither causal thinking nor intuitive psychology, physics, or sociality.[5] A brilliant chess program doesn't know it is playing a game called chess or that its opponent is a human, and it doesn't enjoy the thrill of winning. These differences between human and machine intelligence become clearer when looking at concrete examples. Let's begin by looking at translating language, and then at identifying objects and recognizing scenarios.

Language Translation

In Douglas Adams's *The Hitchhiker's Guide to the Galaxy*, all people need to do to understand a foreign language is insert a small fish into their ear.

Down on humble earth, there are no fish that do the job; instead we need translators. When transposing a text from one language into another, a translator needs to understand the source language while having an even better grasp of the target language in order to express idioms and irony. This is why professional translators typically translate into their mother tongue.

What is the best way to build a translation machine? If you followed the psychological AI program, you would assemble professional translators and linguists in a room and try to convert their intuition and judgment into rules that can be programmed into software. But that has not worked by itself. Language is not a system of well-defined rules: words have not one but many meanings, and the correct one cannot be simply looked up in a dictionary but must be inferred from the context and what one knows about the speaker. Similarly, grammatical rules are not absolute but constantly broken. The alternative is to forget about the art of human translation and instead employ software engineers who use massive computer power to analyze the statistical associations between words and sentences in billions of pages of text already translated. Rules and statistics have been the major two ideas about how to build translation machines.

Until the end of the 1980s, not much progress was made in machine translation. The ALPAC Report commissioned by US funding agencies in the mid-1960s described translation research as flawed and useless.[6] Popular stories made fun of this state of affairs. Here is one. A computer program translated a headline from English into another language and then back into English:

Headline: POPE SHOT. WORLD SHAKEN.

Back translation: EARTHQUAKE IN ITALY. ONE DEAD.

Clearly, the algorithm did "think" something. More correctly, it made associations. I tried to replicate this result in 2020 with Google Translate, a much better system than in the old days. After translating the headline back and forth into German and repeating the process, I got this result:

Headline: POPE SHOT. WORLD SHAKEN.

Back Translation: POPE FIRED. WORLD SHAKEN.

The algorithm struggled with an ambiguous word. *Shot* has two meanings in English: that the Pope *was shot by someone*, or the Pope *shot or fired at someone or something.* Unlike English, German has two different terms for *shot*, one for being shot (*erschossen*) and one for having shot (*schoss*). Thus, when translating the headline into German, the algorithm has to choose the proper term. That requires an understanding of the cause: the world is shaken because someone shot the Pope, not because the Pope shot. People with common sense intuitively know what is meant: that the Pope being shot is the reason that the world is shaken. But modern machine translation systems, from Google to Bing, work by making associations in the absence of causal understanding. That the English term *fired* also has more than one meaning adds to the beauty of the back translation.

The big advantages of state-of-the-art computerized translation systems are their quantity and speed. Good systems can translate texts from over a hundred languages in a few moments, a breakneck pace and an impressive multilingualism beyond human capacity. Understanding and quality, by contrast, are not their strength.[7] Nor is common sense. A computer would translate the sentence "My goldfish barked at the dog" without a blink; a human translator would be startled. To be startled requires understanding a causal relation, namely, that goldfish do not bark. Yet a neural network does not even know that words stand for things. It relates words to words, not to ideas. Therefore, its output ranges from remarkably accurate translations to surprisingly grotesque errors.

Ambiguity and Polysemy Require Common Sense

Lack of common sense is a key problem of all translation systems. For instance, the highly praised system DeepL translated *Pope shot* into German as *Papstschuss*, which was retranslated into *papal shot* (with alternatives such as *shot in the pope* and *shot in the pope's eye*).[8] These systems try to identify entire sequences of words or whole sentences that need to be translated together rather than translating word by word with the help of a dictionary, as in the past.[9] But the quality of the result also depends on the quality of the sources: if many poor translations are on the internet for a particular sentence or topic, then DeepL's suggestions are of poor quality as well. Although translation programs are quite good today, that does not mean that they have common sense. Without understanding, even a good translation system remains an idiot savant.

In contrast to logical languages, natural languages have multiple sources of uncertainty. One is polysemy, the fact that one and the same word may have several meanings—like the English word *shot*. A related kind of uncertainty is a sentence whose meaning cannot be inferred from the individual words but which requires common sense.[10] Here is a classic example:

> Little John was looking for his toy box. Finally, he found it. The box was in the pen. John was very happy.

There is nothing in these sentences that enables the reader to infer the meaning of *pen*. The most frequent one is a writing utensil, but here it refers to a small enclosure in which a child plays, which is less common. I asked DeepL to translate *The box was in the pen* into German, which has different words for the two different meanings. DeepL got it right: *Die Schachtel war im Pferch*—or almost right, since *Pferch* indicates an enclosure for sheep, not children. When I changed one word, *The box was in his pen*, DeepL missed the point, translating it as *Die Schachtel war in seiner Feder*, that is, that the box was in his writing utensil. How to teach a neural network to acquire common sense remains an unsurpassed challenge.

A final source of uncertainty lies in the different grammars of languages. For instance, in German, Italian, Spanish, Polish, and French, the words for male and female professionals differ, while in English they are the same. To translate *nurse* and *doctor* from English to German, the AI has to make a choice between male and female, which requires understanding the context of the story, an easy task for common sense but difficult for the best programs. Even when translating directly from German to Italian, where gender is marked in both cases, Google Translate makes systematic errors in keeping with gender stereotypes. For instance, the German *die Präsidentin* (the female president) is translated as *il presidente* (the male president) in Italian, when it should be *la presidente*. In an experiment where terms were translated between these five languages, Google Translate changed all female doctors into male doctors, all female historians into male historians, and most male nurses into female nurses.[11]

Why does the AI modify the gender to fit the common stereotype? The answer is that Google Translate is English-centered. It does not directly translate from German to Italian, or from Spanish to French, but from each of these languages first to English, and then from English to the target language. In the first step, the gender gets lost in translation, and in the second step Google Translate has to guess it. The best guess is the stereotype.

The Problem Is More General

The fact that AI lacks common sense limits its use not only for translation but for natural language comprehension as well. When people give a reason for a claim, its validity depends not only on the reason but also on "warrants" that are typically left unspoken. An example is:

Claim: You should take an umbrella.

Reason: It is raining.

Warrant: It is not good to get wet.

Warrants are part of humans' intuitive world knowledge. Machines don't have these intuitions. But what if one simplifies the task and provides the warrant explicitly, as above? Can deep neural networks then determine whether a claim is warranted? To test this ability, one can use warrants that support a claim (as above) but also those that do not ("it is good to get wet"). In one study, Google's widely acclaimed neural network BERT (named after the perfectionist character in *Sesame Street*) correctly concluded whether claims are warranted for 77 percent of similar questions. Given that ordinary people without any preparation did only 3 percentage points better, BERT did a truly impressive job.[12]

Yet we should think twice before concluding that the network has learned to comprehend natural language arguments almost as well as people. To do so, it would need common sense and world knowledge—such as that rain gets us wet and that umbrellas protect us from getting soaked. So what did BERT in fact learn to have been so successful? After digging more deeply, the authors of the study discovered the network's ingenious secret: in the set of data on which it was trained and tested, BERT had discovered that a claim was mostly correct when the warrant contained a *not*, as in the example above. By confirming the claim in such cases, it therefore got the answer right most of the time. Yet this remarkable ability to find correlations has nothing to do with actually grasping an argument. When the warrants were reformulated without any *nots*, the network fared no better than chance at giving the right answers. The research team concluded that the neural network's astonishing appearance of comprehending language can be accounted for by its ability to find spurious cues. BERT's top performance is yet another illustration of the Russian tank fallacy.

The Future of Machine Translation

The statistical approach to translation is dominant in computer translation today, often in combination with a rule-based approach. Semantics (the meaning of a word or sentence) and pragmatics (e.g., what a person intends to do) play little role. To understand how these algorithms work, put yourself in the shoes of a recommender system that suggests the next word or phrase to your sentences as you type them. You are a neural network and don't know what a word means. You also know nothing about spelling and grammar. But you have "read" millions of articles and books on all kinds of topics, and you have a perfect memory that recalls all combinations of words and what word or phrase most likely follows a given word or phrase. You "see" the words users enter on their smartphones and predict those that might follow based on those millions of articles, without a clue what these phrases mean. It's similar to a human who does not understand Chinese characters but has learned which one is likely to come after another. The result can be quite impressive, without any understanding of what these words are about.

But when the result is impressive, doesn't that mean that the program understands language? In fact, Ray Kurzweil, who was hired by Larry Page to bring language understanding to Google, argued that statistical analysis is the epitome of understanding: "If understanding language and other phenomena through statistical analysis does not count as true understanding, then humans have no understanding either."[13] This statement confuses the outcome with the process. Even if a program translates a sentence correctly, that does not mean that it comprehends its content, just as a parrot squawking "Ed is a bad boy" doesn't understand what it is saying.

If we could write common sense into a source code, then it would be possible to write it into programs and algorithms. Source codes, however, require a well-defined structure. Language is in a different category—literary texts require understanding of a high level of ambiguity. For that reason, the future of machine translation lies not in the dream of universal automatic translation but in computer-assisted translation systems, including systems that point out inconsistencies in terminology or formatting style within a text or institution. Fast automatic translation will work for well-defined topics that feature logical structures with limited ambiguity, such as news and business texts, and for quick communication where literary quality or creativity is not an issue. It might also work for the needs of the military and intelligence community that funds projects in computer translation to be able to understand the massive amount of foreign language communication they are intercepting. Unlike a competent Arabic translator, a computer can work around the clock and does not need security checks and clearance. Beyond that, one will still need expert translators who understand what the text is about.

Object and Face Recognition

When a little girl says "doggie" while pointing to a dog and "kitty" when pointing to a cat, she recognizes different classes of animals. Object recognition is a fundamental ability; without it, higher-level cognitive abilities such as abstraction, thought, and decision-making would be impossible. But how do toddlers know that an animal is a dog as opposed to a cat? Do they rely on the shape of the eyes, the contour of the head, or the rest of the body? Studies indicate that three-month-old infants already respond differently to dogs and cats and rely on the features of the face as well as the contour of the head, whereas bodily features do not seem to play much of a role.[14] Precisely what cues are used and how these are integrated is not yet known.

The number of instances children need to learn a category is much smaller than for a deep neural network. The latter needs to see tens of thousands of pictures of dogs and cats in supervised learning. A child may need to see a kitty once or a few times and then will recognize cats under different lighting conditions, in the dark or in sunlight. If a three-year-old watches a bike on a road and is told that this is a bicycle, it is likely that from then on, the child will recognize all varieties of bicycles. The son of a colleague of mine became a car enthusiast at age two and could point out all varieties of BMWs on the street, even a model he had not seen before. Children do not appear to be born with this one-shot learning ability but learn it in the first thirty months of life.

Infants Recognize Their Mother's Face
For a computer, recognizing faces is no different a task than recognizing cars. For an infant, these are not the same. Just two days after birth,

newborns can already tell their mother's face apart from a stranger's. They also suck more strongly when they see their mother's face in a video.[15] Recognizing their mother's face is apparently tied to recognizing their first source of food. But face recognition does not arrive full-blown at birth. It takes about ten years for children to develop the face-recognition ability of adults. On the way, they also lose a remarkable skill. Infants and young children can often recognize faces both right side up and upside down, which few adults are still able to do.

The Brain's Vicarious Functioning

The human perceptual system is designed to recognize objects under changing illuminations, situations, and contexts. It is extraordinarily adapted to a constantly changing world. To deal with this uncertainty, the brain uses not one but multiple routes. If one route is blocked, it takes another. This immense flexibility of the brain to rely on changing cues, according to whatever is available, is known as *vicarious functioning*.[16] It can be found in most biological systems. For instance, migrating birds fly thousands of kilometers and may rely on the stars to navigate. But if it's a cloudy day, they may instead rely on landmarks or their magnetic sense. Similarly, the brain can recognize a face by the shape of the eyes, nose, and mouth. In the extreme, only a few lines are sufficient, as when we recognize the face of a celebrity in a caricature made with just a few pencil strokes. Face recognition also functions even when parts of a face are covered and internal features such as eyes are hardly visible. In this case, the brain relies on external features such as hair and head shape.

Consider the image of former US president Bill Clinton and his vice president Al Gore (figure 5.1). How do we recognize Gore on the left and Clinton on the right? Is it Clinton's characteristic nose and mouth? Not at all. In the picture, Clinton and Gore have exactly the same nose, mouth, and eyes. Their entire faces are identical after digital manipulation.[17] Nevertheless, we perceive them as being different. The only real difference lies in the hair and head shape. This shows that if the internal (facial) features are not conclusive, the brain relies on external features—vicarious functioning.

One might call the fact that we see Clinton and Gore rather than two identical faces with different hair and head shapes a visual illusion. Yet dismissing it as such would mean overlooking that the brain, faced with

Figure 5.1
Bill Clinton and Al Gore. Really? Their faces have been edited so that they are identical—eyes, nose, mouth, and all other internal features. Nevertheless, we recognize Clinton and Gore. Our brains identify the two persons by external features, here the hair and head shape. Reprinted by permission from Springer Nature: Sinha and Poggio, "I Think I Know That Face."

uncertainty, has to try different routes to find out who the people in the picture actually are.

A second general feature in human evolution is that faces need to be recognized from a distance to determine whether they are friends or foes. The brain has to compensate for the human eye, which is unable to see distant objects as clearly as, say, an eagle. This requires successfully dealing with blurred or degraded images. Take the picture in figure 5.2, which was obtained by reducing the photo of a famous person to 19 × 25 pixels ("blocks") of various shades of gray. If you look closely, you will hardly recognize the face. But by stepping back and looking at the picture from

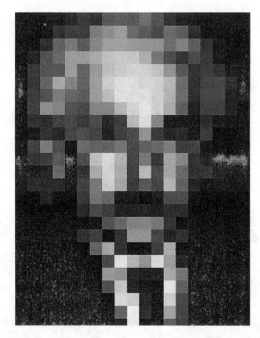

Figure 5.2
Faces need to be identified from a distance to decide whether they are friends or foes, which requires recognizing blurred and degraded pictures. If you step back and look at the picture from a distance, a famous face can be recognized. The same effect can be obtained by deliberately blurring the image, such as by squinting. (Reprinted by permission from Elsevier: Sinha et al., "Face Recognition by Humans.")

a distance, you can. Stepping back can be replaced by squinting, which blurs the picture and fosters recognition. These are humans' adaptations to low resolution, here few pixels, whereas machine perception generally improves when there are more pixels.

Human Errors Differ from Machine Errors

If a system is an improvement on another one, it makes fewer errors—but the errors are typically similar in kind. However, if two systems differ in fundamental properties, like carbon and silicon, these are likely to make qualitatively different errors. Therefore, if artificial neural networks resemble human intelligence, the errors they make should differ in terms of quantity. In fact, they differ in quality. A qualitative difference is an AI error that is unexpected and nonintuitive for a human, or a human error that an AI would never make. We have seen first traces of such errors in the previous chapter with neural networks that recognize school buses.

Human errors that would be baffling for an AI include calculation errors and those due to social influence. Most of us are slow and not very good at mental arithmetic, while computers can make fast and perfect calculations. Moreover, we are social beings who depend on others, and what we believe is influenced by what our friends and family rightly or wrongly believe. Social influence can even affect total strangers. For instance, in the classic conformity experiment, groups of eight students were asked to judge the length of lines drawn on paper.[18] Unknown to the single true participant in each group, the other seven were actors. In some of the trials, the actors gave their best judgments. In a few trials, however, they all gave the same wrong answer. After hearing these, a substantial number of the true participants also gave similar answers. Note that the experiment used a topic as neutral as possible, the length of lines, not political judgments or social issues that lead to group pressure. Even in this situation, a desire for conformity can make many humans err. A computer program couldn't care less what other computers say.

Let's have a closer look at the difference between the errors humans and deep neural networks make.

Counterintuitive Errors

Consider a network that has learned to recognize handwritten numbers through a sample of tens of thousands of images. To simplify the task, each number is set within the boundaries of a box. Then the network is tested with a sample of new handwritten numbers, 0, 1, 4, and 0, as shown in the first column in the left panel of figure 5.3. It classifies the four new samples correctly. Now the interesting part comes. Take a close look at the second column, and you will recognize the same four numbers. But these handwritten numbers are slightly distorted by changes to a small proportion of the pixels, so subtle that the human eye can barely notice any difference. To us, the second column looks like the first; we still see 0, 1, 4, and 0. For a neural network, it's an entirely different matter. It no longer recognizes a single one of the four numbers.[19] The result is the same for the next two pairs of columns.

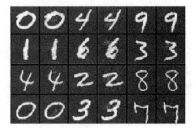 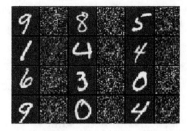

Figure 5.3
Deep neural networks and humans make different kinds of errors. Left panel: The first column contains four handwritten numbers, which a trained deep neural network correctly recognized. The second column is obtained from the first by *systematically* changing a small proportion of the pixels. Although this change is barely detectable by a human, the AI could no longer identify any of the four numbers correctly. The same holds for the next two pairs of columns. Right panel: The first column lists four handwritten numbers; the second shows the same numbers but with a large amount of random noise. The deep neural network could still recognize about half of the numbers, while humans can hardly recognize any of them. (From Szegedy et al., "Intriguing Properties of Neural Networks.")

To us, it is hard to fathom that a well-trained deep neural network doesn't recognize that the numbers in the first two columns are the same. The reason for its confusion is that the pixels have been changed in a tiny but systematic way. When the numbers are distorted by randomly changed pixels (known as *random noise*), however, the network is not fooled in the same way. The first column in the right panel of figure 5.3 shows the numbers 9, 1, 6, and 9, as does the second column but with so much noise added that the human eye can barely make out anything. To us, the second column does not look like the first. The network, in contrast, is still able to identify the barely visible numbers correctly in half of the cases. The network can deal better than humans with noise (right panel), but humans can deal better with tiny systematic changes (left panel). Noise is like a picture covered with dust. Slight systematic changes, in contrast, change the picture, but the brain understands that these are irrelevant and barely perceives them. In contrast, the network appears to focus on aspects that are irrelevant for single digits or is confused by those, reminiscent of the Russian tank fallacy.

The lesson is that the artificial intelligence of neural networks is quite different from human intelligence. Psychological AI would teach the handwriting program to divide the task into subtasks, such as identifying whether there is a horizontal bar or a closed oval shape. That psychological method would not be fooled by systematic changes in irrelevant pixels, as in the second column of figure 5.3 (left). The neural network does not break down that task into such subtasks that are intuitively intelligible to humans.[20] Moreover, when learning novel handwritten digits from 0 to 9, machine-learning algorithms may be trained on a standard set of 60,000 images before being able to correctly recognize new handwritten examples as readily as humans would. Humans, in contrast, can learn to recognize a new handwritten character from just a few examples, learn the associated concept, and generate new examples themselves.[21] Even for simple objects like handwritten characters, people need fewer examples than the best algorithms and learn in a more general way.

What Do Deep Neural Networks See?

We do not know exactly what a network "sees," but we can make some educated guesses. Recall the situation from the previous chapter where networks confused horizontal yellow-and-black stripes with a school bus. In figure 5.4, these two images are placed above that of an electric guitar. After learning to classify the picture on the left as a guitar, the network is tested on the right picture with vertical undulating stripes. Here the network is 99.9 percent confident that this picture is also of a guitar.[22]

Why on earth does the network think that wavy stripes are a guitar? Again, we cannot know for sure, but the juxtaposition of the two errors might give us a clue. In both cases, the misclassified picture has the same dominant colors as the correct object. These are yellow and black for the bus, and brown and white for the guitar. Most guitars are made of wood and therefore brown in color. The body of a guitar is curved like an S on its left side and like an inverse S on its right side. The wave pattern in the picture on the right captures both forms and repeats them in each vertical wave. At least in these cases, the network appears to identify characteristic colors and shapes while learning from pictures of real buses and guitars. And if these colors and shapes are repeated in another picture, the network is highly confident that this image belongs to the same class of objects.

There is no question that deep neural networks have demonstrated impressive abilities in recognizing objects, often on par with humans. A similar level of ability, however, does not mean a similar kind of intelligence.

Figure 5.4

Top row: As shown in figure 5.3, the picture on the left is correctly classified as a school bus by a deep neural net, but the picture on the right is also classified as a school bus with high confidence. Bottom row: The neural network has learned to correctly classify the picture on the left as a guitar, but is 99.9 percent certain that the picture on the right is also a guitar. Based on Szegedy et al., "Intriguing Properties of Neural Networks," and Nguyen et al., "Deep Neural Networks." © IEEE. Reprinted with permission from Nguyen et al., "Deep Neural Networks."

A woman riding a horse on a dirt road	An airplane is parked on the tarmac at an airport	A group of people standing on top of a beach

Figure 5.5
Image captions generated by a deep neural network. The network mostly gets the objects right but fails to understand the relations between the objects, the mental states of the people, and the physical forces at work. Image credits, from left to right: Gabriel Villena Fernández/Wikimedia Commons, picture alliance/AP Images, and picture alliance/AP Photo/Dave Martin. Similar images can be found at twitter.com /interesting_jpg. From Lake et al., "Ingredients of Intelligence."

The systematic errors made by neural nets are alien to humans. The networks can be fooled because they appear to rely on recurrent features without comprehending what the object is.

Scenario Recognition

Recognizing objects is relatively simple compared with recognizing relations between persons, objects, or states. Here, the question is: What is going on in a scenario? Humans infer the answer with the help of their intuitive psychology and physics. How do algorithms compare?

Let's look at how deep neural networks trained to generate image captions fared at recognizing causal relationships (figure 5.5).[23] The image on the left shows a take from a cowboy stunt show at the Texas Hollywood theme park. It depicts a violent scene from a typical old Western movie where an outlaw is lassoed and dragged behind a horse. In the background, visitors are watching the performance. What does the network "see"? The neural network's caption is "a woman riding a horse on a dirt road." The network gets the objects mostly right—the horse and the dirt (the person on the horse is a man, which is hard to see, although that could be inferred from the standard plots of Westerns). What the caption reveals, however, is that the network doesn't have a clue what is going on in the scene. It has no intuitive psychology to infer that the person on horseback intends to punish the other person, and that the scene is part of a show. It also has no causal understanding that this kind of punishment may be fatal.

The center image depicts the crash of a passenger plane in Taipei, on February 2015, filmed by a car driver. Shortly after takeoff, the right engine had a malfunction, and the pilots mistakenly shut down the still-functioning left engine. In the voice recording, one of the pilots shouted, "Wow, pulled back the wrong side throttle." The plane rolled sharply, struck a taxi with its left wing, and toppled into the river.[24] Forty-three passengers and crew members were killed. The neural network's caption is "an airplane parked on the tarmac at an airport." Once again, it gets some of the objects right but misses the story.

The third image shows residents in Key West, Florida, clinging to each other as they battle the winds of Hurricane Georges along Houseboat Row in 1998. Here, the deep neural network's caption is "a group of people standing on top of a beach." It cannot make heads or tails of what is happening.

When humans look at these pictures, they rely on their common sense and experience to guess what is going on. Even if we've never seen a Western movie, a plane crash, or a hurricane, our intuition tells us that something is awry. Deep neural networks can be quite good at identifying objects in a picture but, without the ingredient of common sense, struggle to understand how they relate in a scenario.

Of Different Minds

Human intelligence is about representing the world, making causal models, and ascribing intentions to other living creatures. To do this, humans distinguish an image from the real thing, such as knowing that the picture of a person is not the person, although similar emotions can sometimes be elicited by both. In contrast, deep neural networks learn to associate images with labels or headlines but do not know that an image refers to some person or object in the real world. AlphaGo and its successors play Go better than human champions without knowing that they are playing a game, and digital assistants like Siri and Alexa don't know what a restaurant is. Does it matter that they don't know?

As long as the assistant is only asked about the best Italian restaurants in your neighborhood or similar advice, it does not. Yet awareness is extremely important when systems are allowed to automatically make decisions with life-and-death consequences, as with military drones, robot soldiers, and other lethal autonomous weapons. A machine may know perfectly how to kill, but it does not know what it is doing and why. What is more, these machines can err in unexpected ways that are alien to our imagination.

The human person that reminds me most of the performance of a deep neural network is Solomon Shereshevsky, the famous Russian mnemonist, with whom readers of my book *Gut Feelings* are already acquainted. His memory appeared to have no limits in its capacity and duration.[25] When asked to read a page of text, Shereshevsky could recall it word for word, both forward and backward. But when asked to summarize the gist of what he had read, he was more or less at a loss. He had problems with ambiguity, with words that have several meanings and with different words that have the same meaning, not to mention metaphors and poetry. Shereshevsky could precisely recall a complex mathematical formula even though he could not understand it (to be sure, it was made up), and also recall it fifteen years later. His mind was very different from that of chess masters who can also perfectly recall complex chess positions but only if they make sense, not random configurations. Shereshevsky struggled with filtering the important from the trivial and reasoning on an abstract level.

Shereshevsky is an existence proof that evolution could have given all of us perfect memory but at a steep price. One thing he could not do was forget. Although his vast memory seems enviable, he was distracted by irrelevant details, not unlike deep neural networks that are distracted by irrelevant pixels added to handwritten numbers, as in figure 5.3, or by a patch on people's T-shirts, as in figure 4.4. He may have been the closest humans can get to neural networks, perfect in storing and processing big data but having trouble understanding what it all actually means.

6 One Data Point Can Beat Big Data

Out with every theory of human behavior, from linguistics to sociology. Forget taxonomy, ontology, and psychology. . . . With enough data, the numbers speak for themselves.
—Chris Anderson, "End of Theory"

There are a lot of small data problems that occur in big data. They don't disappear because you've got lots of stuff. They get worse.
—Sir David J. Spiegelhalter, quoted in Harford, "Big Data"

Astronomy is the science that for centuries has required its researchers to work at night. It is also the science that initiated one of the first big data projects. Launched in 1887 in Paris under the name *Carte du Ciel* (sky map), the project set out to map two million stars using 20,000 photographic plates of the night sky, documented in hundreds of volumes of published data.[1] This undertaking fit the newly created term *big science*, as it gobbled up almost all of the limited resources available to the observatories and the labor and time of generations. Mapping the stars in all hemispheres required international cooperation, using observatories from Helsinki to the Cape of Good Hope to Sydney. The sky map promised to deliver enough images and data to transform astronomy into a day job where astronomers could sit at a well-lit desk rather than out in dark fields. Most remarkably, it was intended not to promote the fame of individual astronomers but instead to provide a service to future "astronomers of year 3000 at least."[2] With the help of the photographic plates, future scientists would be able to detect small changes in the sky that are not noticeable within the lifetime of an astronomer.

The unfinished sky map project was a monument to *positivism*. The term refers to an attitude that what counts are facts, that is, everything that can be observed and measured, as opposed to imperceptible ideas and speculation. Today, positivism has made a comeback with big data analytics.

Astronomers deal with a stable system: the movement of heavenly bodies. The system is stable relative to the short duration of astronomers' lives. Unlike the typical machine-learning application, astronomy has theories about stars and planets. In this context, big data is highly useful. In today's world, however, big data is used for fickle phenomena that are dynamic and may change in unexpected ways. Here, the three Vs—volume, velocity, and variety—are of limited help, and less can be more: using fewer data and less complex algorithms can often lead to better predictions. Even adding a fourth V—veracity, that is, the reliability of the data—can be of little avail. In its place, psychological AI can be more useful. Let me illustrate this less-is-more principle with Google's celebrated showcase of big data analytics.

Predicting the Flu

If you are experiencing high fever, a sore throat, a runny nose, and tiredness, you might have influenza, commonly known as the flu. The symptoms appear typically two days after being exposed to the influenza virus and disappear within five to six days. Influenza is estimated to cause a quarter to half a million deaths around the world every year. To indicate where the flu is spreading, the US Centers for Disease Control (CDC) informs the public about the number of flu-related doctor visits in all regions of the US. The problem is that it takes a week or two for the CDC to collect the data.

In 2008, media across the world announced with fanfare that Google engineers had found a much quicker method to predict the spread of the flu early on. The idea appeared sound. Users infected with the flu are likely to use Google's search engine to diagnose their symptoms and look for remedies. These queries could instantly tell where the flu is spreading. To find the apt queries, engineers analyzed some fifty million search terms and calculated which of these were associated with the flu.[3] Then they tested 450 million different models to find the one that best matched with the data and came up with a secret algorithm that used forty-five search terms (also kept secret). The algorithm was then used to predict flu-related doctor visits in each region on a daily and weekly basis.

At first, all went splendidly. Google Flu Trends forecasted the flu faster than the reports of the Centers for Disease Control. Google even coined a new term: to "nowcast" the spread of flu and influenza-related diseases in each region of the United States, with a reporting lag of about one day.

Months later, in the spring of 2009, something unexpected happened. The swine flu broke out. It barreled in out of season, with the first cases in March and a peak in October. Google Flu Trends missed the outbreak (see figure 6.1); it had learned from the previous years that flu infections were high in the winter and low in the summer.[4] Predictions crumbled.

Faith in Complexity

After this setback, the engineers embarked on improving the algorithm. To do so, there are two possible approaches. One is to fight complexity with complexity. The idea is that complex problems need complex solutions, and if a complex algorithm fails, it needs to be made more complex. The second approach follows the stable-world principle. The idea behind it is that a complex algorithm using big data from the past may not predict the future well in uncertain conditions, and it therefore should be simplified. Google's engineers went for more complexity. Instead of paring down the forty-five search terms (features), they jacked these up to about 160 (the exact number has not been made public) and continued to bet on big data.

The revised algorithm did a good job at first of predicting new cases, but not for long. Between August 2011 and September 2013, it overestimated the proportion of expected flu-related doctor visits in 100 out of 108 weeks (see figure 6.1).[5] One major reason was the instability of the flu itself. Influenza viruses are like chameleons, constantly changing, making it extremely difficult to predict their spread. The symptoms of swine flu, such as diarrhea, differed from those in past years, and the infection rate was higher for younger people than with other strains of the flu. A second reason was the instability of human behavior. Many users entered flu-related search terms out of sheer curiosity about swine flu, not because they felt sick. But the algorithm could not distinguish between the motivations for search. The engineers asked, "Is our model too simple?" and continued to tinker with the revised algorithm, to no avail.[6] In 2015, Google Flu Trends was quietly shut down.[7]

Some may shrug and say, "Yes, we've heard this all before, but that was 2015; today's algorithms are infinitely bigger and better." But my point is not the success or failure of a particular algorithm developed by Google. The crux is that the stable-world principle applies to *all* algorithms that use the past to predict an indeterminable future. Before Google's big data analytics flopped, its claim to fame was taken as proof that scientific method and theory were on the brink of becoming obsolete. Blind and rapid search through terabytes of data would be sufficient to predict epidemics. Similar claims were made by others for unraveling the secrets of the human genome, of cancer, and of diabetes. Forget science; just increase volume, velocity, and variety and measure what correlates with what. Chris Anderson, editor-in-chief of *Wired*, announced: "Correlation supersedes causation, and science can advance even without coherent models. . . . It's time to ask: What can science learn from Google?"[8]

Let me pose a different question: What can Google learn from science?

Under Uncertainty, Keep It Simple and Don't Bet on the Past

The Google engineers never seem to have considered a simple algorithm in place of their big data analytics. In my research group at the Max Planck Institute for Human Development, we've studied simple algorithms (*heuristics*) that perform well under volatile conditions. One way to derive these rules is to rely on psychological AI: to investigate how the human brain deals with situations of disruption and change. Back in the early nineteenth century, for instance, Thomas Brown formulated the Law of Recency, which states that recent experiences come to mind faster than those in the distant past and are often the sole information that guides human decision.[9] Contemporary research indicates that people do not automatically rely on what they recently experienced; they only do so in unstable situations where the distant past is not a reliable guide for the future. In this spirit, my colleagues and I developed and tested the following "brain algorithm":

> *Recency heuristic for predicting the flu*: Predict that this week's proportion of flu-related doctor visits will equal those of the most recent data, from one week ago.[10]

Unlike Google's secret Flu Trends algorithm, this rule is transparent and can be easily applied by everyone. Its logic can be understood. It relies on

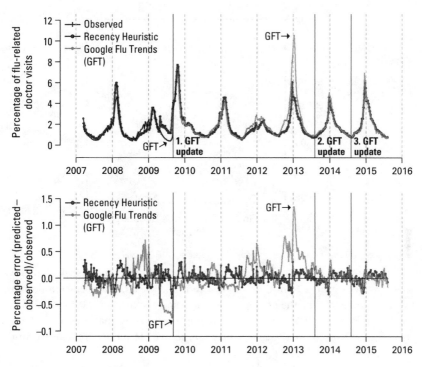

Figure 6.1

A simple heuristic using a single data point can predict the flu better than Google's big data analytics. Shown here is the actual percentage of flu-related doctor visits from March 18, 2007, to August 9, 2015, and its predictions by the recency heuristic and by Google Flu Trends (including three updates). Top: Predictions and observed values in absolute terms. The predictions by the recency heuristic and the observed values are virtually identical. Bottom: Prediction errors. The years signify the beginning of the year, that is, *2008* indicates January 1, 2008. For instance, in the summer of 2009, Google Flu Trends underestimated the spread of the flu thanks to the unexpected breakout of the swine flu, after which it received its first update. Source: Katsikopoulos et al., "Transparent Modeling."

a single data point only, which can be looked up on the website of the Centers for Disease Control. And it dispenses with combing through fifty million search terms and trial-and-error testing of millions of models. But how well does it actually predict the flu?

Three researchers and I tested the recency heuristic using the same eight years of data on which the Google Flu Trends algorithm was tested, that is, weekly observations between March 2007 and August 2015. During that

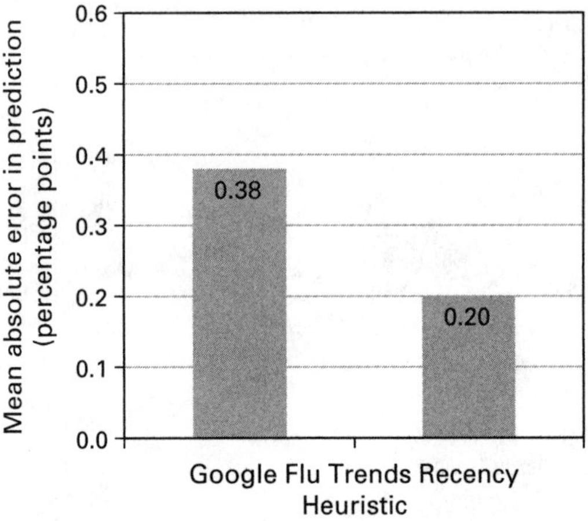

Figure 6.2
Less can be more. Using a single data point, one can predict the spread of the flu better than Google Flu Trends, a big data algorithm. The mean absolute error (from figure 6.1) in predicting the proportion of flu-related doctor visits is 0.38 for Google Flu Trends, but only 0.20 when using one single data point, that is, recency. Both algorithms were tested on the same weekly data between March 18, 2007, and August 9, 2015.

time, the proportion of flu-related visits among all doctor visits ranged between 1 percent and 8 percent, with an average of 1.8 percent visits per week (figure 6.1). This means that if every week you were to make the simple but false prediction that there are zero flu-related doctor visits, you would have a mean absolute error of 1.8 percentage points over eight years. Google Flu Trends predicted much better than that, with a mean error of 0.38 percentage points (figure 6.2). The recency heuristic had a mean error of only 0.20 percentage points, which is even better.[11] If we exclude the period when the swine flu happened, that is, before the first update of Google Flu Trends, the result remains essentially the same (0.38 and 0.19, respectively).

"Fast-and-Frugal" Psychological AI

The case of Google Flu Trends demonstrates that in an unstable world, reducing the amount of data and complexity can lead to more accurate predictions. In some cases, it might be advisable to ignore everything that happened in the past and instead rely on the most recent data point alone. It also shows that psychological AI—here, the recency heuristic—can match or beat complex machine-learning algorithms in prediction. In general, my point is that "fast-and-frugal" heuristics that need little data are a good candidate for implementing psychological AI.

The flu example is neither a fluke nor an exception. Under uncertainty, simple rules such as recency have also been shown to be highly effective in comparison with complex algorithms, be it in predicting consumer purchases, repeat offenders, heart attacks, sports results, or election outcomes.[12] A group of economists, including Nobel laureate Joseph Stiglitz, showed for instance that the recency heuristic can predict consumer demand in evolving economies better than traditional "sophisticated" models.[13] The great advantage of simple rules is that they are understandable and are easy to use.

Nevertheless, it's difficult for many of us to get around the idea of deliberately leaving out data when we're trying to make an informed decision. Why is having more information often a hindrance rather than a help?

As mentioned in chapter 2, to successfully predict the future, one needs a good theory, reliable data, and a stable world. The usefulness of big data depends on these three conditions. Let's first take a look at situations where one has lots of correlations but no theory.

Correlations and the Texas Sharpshooter

Although not as popular as the Oscars, the Nobel Prize is one of the most prestigious international awards, and its winners make headlines each year. The United States has about ten Nobel Prize winners per ten million inhabitants, and the UK almost twice as many. China and Brazil are at the low end, with less than 0.1 winner per ten million, while Switzerland and Sweden are at the top, with over thirty Nobel laureates per ten million. What causes these differences? And what should scientists and writers from other countries do to match the level of Switzerland?

The answer of big data is to find out what individual behaviors or organizational structures are associated with the proportion of Nobel laureates. It could be the quality of the preschool system, or that of the universities, or the motivation to succeed. If one indiscriminately combs the data in search of an association, an astonishing answer pops up.

It's chocolate! The rate of Nobel laureates in a country can be "predicted" by its chocolate consumption. The more chocolate eaten, the more Nobel

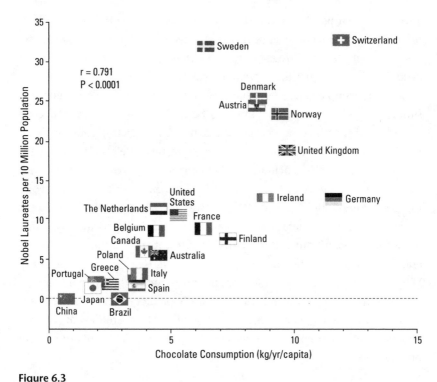

Figure 6.3
An impressive but useless correlation. Chocolate consumption is strongly correlated
($r = 0.79$) with the number of Nobel laureates per ten million population across countries. The value $p < 0.0001$ means that if there were no correlation in reality, a correlation at least as strong as this one would be expected to occur only once in 10,000
cases, or even less frequently. From Messerli, "Chocolate Consumption." Reprinted
with permission from Massachusetts Medical Society.

Prizes (figure 6.3). And the relation is extremely strong. The Chinese and
Japanese consume very little chocolate and have very few Nobel laureates
per capita. The Swiss consume on average more than twenty-six pounds of
chocolate per year and are at the top of the Nobel laureate scale. The United
States is somewhere in the middle in both respects. (One exception to the
rule is Germany, whose citizens consume as much chocolate as the Swiss
yet have fewer Noble Prizes per capita; another exception is Sweden, whose
citizens get their high share of Noble Prizes despite eating only an average
dose of chocolate.)

The recommendation seems to be to modify the Chinese, Japanese,
and US diet to one that contains substantially more chocolate, preferably

Swiss chocolate. That may provide chocoholics an alibi for upping their chocolate intake but will not likely bring them any nearer to winning a Noble Prize.

Impressive But Useless Correlations

The r that measures the association in figure 6.3 is called a *Pearson correlation*. Karl Pearson, born in 1857 into a Quaker family of stern, industrious Yorkshire barristers, was a man whose profound curiosity about the world led him to study as much as he could: mathematics, physics, physiology, history, Roman law, and German literature in Cambridge, Berlin, Heidelberg, and Vienna. He also was a man of profound self-doubt, worrying that his name would survive merely as that of a correlation coefficient. At a time when women were not allowed to vote, he argued for the equality of men and women and a strong independent women's party. More relevant to the present topic, Pearson worshipped quantification and argued that our perceptions are the basis of all knowledge: we can perceive correlations but not causation.[14] He was not the first to claim this. The Scottish philosopher David Hume famously argued the same 150 years earlier. As the first epigraph to this chapter shows, big data enthusiasts have pushed Pearson's and Hume's argument to the extreme, arguing that causes are not even needed; rather, with petabytes in our hands, "correlation is enough."[15]

The strong association between chocolate consumption and Nobel laureates shows that correlation is not enough. Data mining may even lead us on the wrong trail. Big data can unearth any number of similar useless correlations. When developing Google Flu Trends, for instance, the engineers found a strong correlation between searches for "high school basketball" and flu-related doctor visits. No causal connection exists, only the coincidence that both the flu and the high school basketball season last from November to March.[16] Applying their common sense, the engineers eliminated this irrelevant feature by hand.

Blind search through millions of variables can reveal many other surprising correlations (figure 6.4).[17] Every year, around a hundred Americans fall into swimming pools and drown. Why? Data mining shows that this number is strongly associated with the number of movies in which the actor Nicolas Cage appears. As you can see in the figure, these numbers closely follow each other over a period of ten years. The correlation between drowning and the number of Nicolas Cage films is 0.67, which

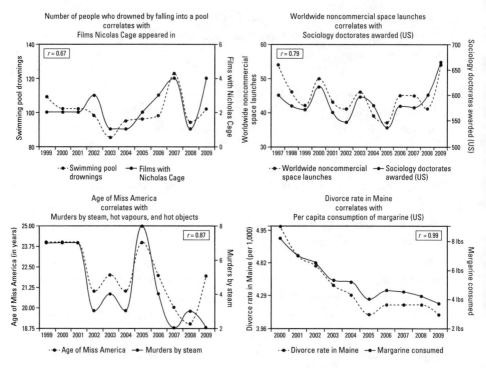

Figure 6.4
If you collect enough data, you can find whatever you want. A mixed bag of non-sense correlations obtained from blind search. Source: TylerVigen.com/old-version.html.

amounts to an unusually high correlation in the social sciences (a value of 1 is a perfect correlation, a value of 0 means no correlation). Do more people really drown because the actor has appeared in more films? Likely not. If you search the gigantic amount of data available on all actors and all causes of death, you'll likely find more of these associations, at least within some time interval, even if they make no sense.

Similarly, a correlation of 0.79, as strong as between chocolate and Nobel laureates, has been reported between the number of noncommercial space launches and the sociology doctorates awarded in the United States (figure 6.4). Granting more PhDs in sociology, then, promises to be a thrifty way to launch more space shuttles.

Even more impressive is the correlation of 0.87 between the age of Miss America and the number of people murdered by steam, hot vapors, and hot objects. A correlation of 0.87 is a rare event. When Miss America is twenty years old or younger, fewer murders occur. The older Miss America is, the more people are murdered. Following that logic, the beauty pageant committee makes life-and-death decisions every time it chooses a winner.

None of these findings, however, can compete with the virtually perfect correlation of 0.99 between the divorce rate in Maine and the per capita consumption of margarine in the United States. Recall the attempts of psychologists to predict divorce rates, reported in chapter 2, using the Texas sharpshooter method. They were studying the verbal and nonverbal interaction between husband and wife. What this near-perfect correlation suggests is that the true cause of divorce appears to have escaped the psychologists, namely eating too much margarine. That in turn suggests we can predict divorce rates perfectly. And it also suggests a solution: by striking margarine from their grocery lists, couples would stay together until death do them part. Once again, by indiscriminately searching through *all* food sources and *all* states in the United States, you might also strike it lucky in digging up another perfect correlation. If the divorce rates in other states or within other time intervals were chosen, this perfect correlation would dissolve into hot air.

In all of these cases, common sense tells us that the correlations are meaningless. Although the examples here are comical, the problem becomes serious when it comes to areas such as nutrition and health, where so many possibilities exist that something is sure to correlate with something else. Much of the nutritional advice we've received over the past years stems from spurious correlations found in studies: Eat blueberries to prevent memory loss. Eat bananas to get a higher verbal SAT score. Eat kiwis late at night to sleep better.[18] The common feature is that next year you may well read the opposite advice.

Another serious problem emerges when an agency relies on blind search to make decisions on people's creditworthiness. As the founder of a credit agency proclaimed, "We observed that people who did not repay their loan had a very particular font on their computer."[19]

Millions of Bullet Holes

By now, you may have recognized the method behind these correlations. Big data analytics provides a whole new field for the Texas sharpshooter method. Only here the sharpshooter no longer needs to shoot at the barn first and then paint the target around the bullet holes so that the bull's-eye is in the middle. Instead, millions of bullet holes are already out there, and the sharpshooter can use a fast search algorithm to find a pattern where the holes appear to be aligned before painting a target around them, as in the four cases in figure 6.4. Once again, the blame is not on big data but on how it is being used. You might stumble over an unexpected correlation, but before proclaiming it to be a fact, you would need to independently test whether it holds up at other points in time or with other groups of people. By skipping this step and presenting the coincidental pattern that happened to be found as a scientific result, you would be promoting the sharpshooter fallacy. There are strong incentives to commit this fallacy, including the media's and public's insatiable hunger for breaking news and the desperate hope of those who suffer from an incurable disease of finding a cure.

The incentives to report a finding before it is independently confirmed are particularly strong in health care. Every couple of weeks, a headline appears about a new tumor marker that promises personalized diagnosis or even a cure for cancer. After raising excited hopes, these breakthroughs often prove to be false, again due to spurious correlations. When another study is done, at a different point in time or in a different part of the world, the correlation vanishes into thin air. A team of scientists at the biotech company Amgen tried to replicate the findings of fifty-three "landmark" medical articles. In forty-seven cases they failed. Likewise, the pharmaceutical company Bayer examined sixty-seven findings in oncology, cardiovascular medicine, and women's health and were able to replicate the results in only fourteen cases.[20] Ian Chalmers, one of the founders of Cochrane, and Paul Glasziou, chair of the International Society for Evidence-Based Health Care, estimated that 85 percent of all health research is "wasted" by not being reproducible, resulting in $170 billion in losses worldwide per year.[21] This dismal situation has been dubbed the *replication crisis*. Results cannot be replicated because they were nothing more than spurious correlations.

Mere correlations without theory are one factor that limits the success of big data analytics; unreliable data are another.

Bad Big Data

A *Forbes* journalist received an ad to join an association for retired people. He was puzzled; not only did he not feel like sixty-five, but he was in his mid-thirties.[22] After more ads arrived, he began to realize that a profile of him had been made and sold by data brokers to the association. Data brokers tend to act in the shadow of better-known companies, and many of their products—that is, people like the journalist, you, and me—haven't even heard of Acxiom, Oracle Data Cloud, and the other companies that construct profiles of hundreds of millions of people. These profiles are based on cookies, browsing behavior, and other sources. They have been bought, for instance, by Facebook to learn more about its users.

The journalist contacted both Acxiom and Oracle and eventually got his hands on the profiles. At first glance, they were absurdly off base. He was listed as over sixty-five, married, a frequent shopper at Victoria's Secret, and a consumer of huge amounts of diapers, baby food, and imported beer—none of which was the case. Three-quarters of the information was utterly wrong, and each broker provided different misinformation. When he asked Acxiom where it got these data, the company insisted that this was a trade secret and continually reiterated how accurate their data are. This case may be an anomaly, and the data on the other millions of people may be correct. But when a *Reuters* journalist also asked Acxiom for his profile, he found it inaccurate in similar ways.[23] When he, too, asked why so much was plain wrong, the company explained that most of what was on his profile was inferred by making guesses from other data, such as his postal code and household history. Other reporters who checked their profiles similarly found an alarming level of false information.[24]

These are individual anecdotes, so what does research show? One field study investigated how accurate the services of data brokers are for delivering ads to the right audience.[25] The ads were for a charity campaign whose targets were males between the age of twenty-five and fifty-four (who comprised 27 percent of the male population). Identifying this age group shouldn't be that difficult. Emulating a typical advertising campaign, the researchers used the combined services of data brokers and ad-buying platforms (which "optimize" campaigns and help select the websites on which the ads are placed). Nineteen leading data brokers and six ad-buying platforms were examined. Surprisingly, the average accuracy was a disappointing 59 percent, meaning that 41 percent of the ads were off-target. The best provider showed the ads to the right targets 72 percent of the time, the worst provider only 40 percent. A second study looked at data brokers' ability to identify the same target group alone, that is, without the help of any ad-buying platforms. This time, their accuracy dropped below chance level.

Similarly, simply guessing did a better job than data brokers at identifying gender. The best results were obtained when data brokers identified audience interest, such as interest in sports and fitness (an increase of twenty and thirty-four percentage points, respectively, over guessing). Interests appear to be more easily inferred from browsing behavior. All in all, the quality of the data provided by brokers does not look like the promised "new oil" or "new gold."

Moreover, advertisers have to pay good money for the oil. The authors of the studies concluded that, for standard display banner ads, the additional costs of targeting are often so high that these may outweigh the additional gains.

Big Data and Unstable Worlds

Beyond mere correlations and unreliable data, lack of a stable situation is the third factor that makes big data analytics of limited use. Consider children at risk of failing school and families at risk of losing their jobs and homes. If an algorithm could identify those at risk, authorities could anticipate the problems before they arise and send social workers to work with these families. A team of researchers from Princeton University challenged the scientific community to predict the future of over 4,000 "fragile families," which were mostly single parents or families of unmarried couples.[26] For each family, the life outcomes to be predicted included the child's expected GPA at age fifteen; the child's grit, that is, the ability to work hard and persevere; whether the primary caregiver will lose their job; and whether the family will be evicted from their home for not paying the rent or mortgage. The researchers provided big data collected over fifteen years, including personality tests, biometric data, interviews, and in-home assessments. There were more than 12,000 measures per family, resulting in millions of data points.

One hundred and sixty science teams participated in the challenge, many using complex state-of-the-art machine-learning methods to make their predictions. The result was discouraging. Most of the algorithms did a poorer job of predicting than a simple rule that used four data points only, such as the mother's marital status and the child's performance when they were last tested (six years earlier). The team in charge of the challenge concluded that even the best algorithms predicted poorly and only slightly better than the simple rule.

In the United Kingdom, a similar study was conducted.[27] Thirty-two machine-learning algorithms were tested on whether they could predict which child or young adult was at risk. Not a single one predicted well: on average, they missed four out of five children at risk. And, when an algorithm did conclude that a child was at risk, it was wrong for six out of ten children. That means if family services were to trust these algorithms and send social workers to solve the problem before it arises, they would miss most of the children at risk and spend their time working with the wrong families in the wrong places. Last but not least, families might be stigmatized in the process.

You may be surprised by these findings. Didn't machine learning beat the best chess player years before these studies were made? Perhaps, one might suspect, the scientists did not collect enough data. But there was in fact more than enough information. The answer is instead a very different one. Life is not like chess; it contains plenty of uncertainty. Too many factors determine what happens to a child, adult, or family, and in fragile families the interplay between these factors may even be amplified. Computational power and big data are of limited help when uncertainty reigns.

We hear about highly accurate data and powerful predictions not always because they actually exist, but because these stories need to be told to sell products and stay in business. In many countries, the use of automated decision-making—such as for identification of vulnerable children, predictive policing, face-recognition mass screening—increases despite mounting evidence of its inaccuracy.[28] Its spread is driven instead by economic and psychological factors. The main psychological drivers include a belief in the objectivity of algorithms and the fear of missing out and falling behind other countries or companies. This combination of beliefs and fears pushes us into ever more commercial and governmental surveillance, even if it is riddled with errors.

Do Numbers Speak for Themselves?

Numbers, big or small, cannot speak. Numbers are like children, requiring attention and guidance. They need to be understood. The COVID-19 crisis brought this message home. When, in mid-February 2020, the number of new infections suddenly skyrocketed in China from one day to the next, the world panicked. But the numbers had changed their meaning. China had lowered the criteria for diagnosing people as infected, no longer on the basis of lab test results but of symptoms and CT scans only. That made the number of new infections soar. It also made the numbers before and after no longer comparable. When one Sunday in mid-March 2020, the number of new infections dropped for the first time, Germans hoped that the downward spiral had begun. But the numbers were collected on a daily basis from the local health authorities, which were only partially staffed on weekends and unable to report all numbers. On Monday and Tuesday, numbers rose again, and hopes sunk. Slowly, the public learned that the lower infection numbers that began to appear every weekend were due to the delay in reporting, not to less viral activity. The number of new infections reported on one day was not the actual number of new infections that day.

The most frightening numbers publicized in 2020 were the coronavirus death counts. The COVID-19 dashboard maintained by Johns Hopkins University reported the numbers by country. Yet these numbers were hardly comparable either. Countries and even regions counted deaths in different ways. Italy, which was hit first in Europe, counted people who had died *and* had a positive COVID-19 test result. That means the person may have died *from* the virus or may have died *with* the virus but from another cause. Moreover, the cause of death is notoriously difficult to determine if the deceased had several serious diseases. During the pandemic, the dimensions of this problem became clear when the Italian health organization published the data for the first 45,500 coronavirus-related deaths.[29] Ninety-seven percent of those had one or more preexisting severe diseases, including 11 percent stroke, 17 percent invasive cancer, 22 percent dementia, and 28 percent ischemic heart disease. Only 3 percent had no comorbidities. In addition, half of those who died were between 82 and 109 years old. In many of these cases, the cause of death was not clear—possibly even several causes that worked together.

If all countries used the same definition—death *and* positive test—then it would at least be possible to compare the numbers. Consider Belgium, where the total number of coronavirus-associated deaths was higher than in Germany, despite its much smaller population. It turned out that the Belgian authorities reported COVID-19 deaths for people who had not even been tested, such as people who died in care facilities for the elderly in which another resident tested positive. The general lesson is: numbers need to be understood before they are added and compared. This is true for all data, whether big or small.

So where is big data useful? As mentioned earlier, it is most useful in situations that are stable, where data are reliable, and where theory can guide search. Examples of these situations are astronomy and the analysis of past data, as in health records—unless the algorithms are gamed. Big data is less promising in situations that may change unexpectedly and where one is looking for the needle in a huge haystack of data. Examples of these situations are predicting spread of the flu, currency exchange rates, children at risk, or human behavior in general.[30] In statistics, the law of large numbers describes a situation where having more data is better for making predictions. According to it, the more often an experiment is conducted, the closer the average of the results can be expected to match the true state of the world. For instance, on your first encounter with roulette, you may have beginner's luck after betting on 7. But the more often you repeat this bet, the closer the relative frequency of wins and losses is expected to approach the true chance of winning, meaning that your luck will at some point fizzle out. Similarly, car insurers collect large amounts of data to figure out the chances that drivers will cause accidents, dependent on their age, region, or car brand. Both casinos and insurance industries rely on the law of large numbers to balance individual losses. Yet that works only as long as the situation in which they are operating remains stable.

II High Stakes

The first principle is that you must not fool yourself, and you are the easiest person to fool.

—Richard P. Feynman

7 Transparency

Personal data shall be processed lawfully, fairly and in a transparent manner in relation to the data subject.
—Article 5 (1a), EU General Data Protection Regulation

Everything should be made as simple as possible, but not simpler.
—Attributed to Albert Einstein

Eric Loomis, a Wisconsin man, was arrested after being caught driving a car that had been used in a shooting. Loomis denied having been part of the shooting but pleaded guilty to less severe charges: attempting to flee a traffic officer and driving a car without the owner's consent.[1] To determine his sentence, the judge consulted his criminal history, as usual, but also a risk assessment algorithm, which classified Loomis as at high risk of reoffending. Loomis was sentenced to six years in prison. Neither the defendant nor the judge knew how the algorithm calculated this risk: the algorithm, known as COMPAS (Correctional Offender Managing Profile for Alternative Sanctions), is a business secret. Loomis appealed on the grounds that the judge violated due process by relying on an opaque algorithm that may discriminate against black men. However, the Wisconsin Supreme Court ruled against Loomis, arguing that it would have imposed the same sentence regardless of the algorithm and that secret algorithms do not violate due process, while also recommending skeptical caution in their use. According to a *New York Times* headline, Loomis was "Sent to Prison by a Software Program's Secret Algorithms."[2] That headline may have overstated the case, but it put a general problem on the table.

A *black box algorithm* is an algorithm that is not transparent, because it is either secret or too complicated for users to understand. Varying by country, secret algorithms influence the lives of many citizens by determining decisions about parole, bail, sentencing, social assistance, loans, and creditworthiness in general. Given that they lack transparency, which violates our intuitive understanding of justice and can conflict with due process, wouldn't it be better to ban them entirely from the courtroom? After all, risk scores can be equally obtained from more transparent algorithms, where it is easier to determine whether these are trustworthy and whether they discriminate against certain groups of people, meaning that it is also much easier to improve or correct them if they fail. Defendants and judges alike should not be left in the dark about how a risk score is calculated.

Black Box Justice

Forecasting criminal behavior is difficult, surprisingly difficult. Psychiatrists and health professionals are regularly asked by courts to predict the probability that a defendant will commit a violent act in the long-term future. According to the American Psychiatric Association in a communication to the US Supreme Court, its "best estimate is that two out of three predictions of long-term future violence made by psychiatrists are wrong."[3] When they predict no violence, they are still wrong in one out of ten cases. Despite these sobering numbers, the Supreme Court ruled that such testimony is legally admissible as evidence, noting that the experts "were not always wrong . . . only most of the time."[4] Alongside psychiatrists, judges have also been blamed for being unreliable. An infamous study in Israel made headlines by concluding that judges are literally influenced by their guts. At the beginning of the day, judges granted parole to about two-thirds of the prisoners, a proportion that declined steadily over time to nearly zero. After the judges had a meal or snack break, the next prisoners in line once more had a two-thirds chance of being granted parole, and this chance steadily declined a second time toward zero as the judges grew hungry again.[5] After the next food break, the pattern repeated itself. Yet the authors had overlooked an important point: that the order of prisoners was not random. The court tried to complete all cases from one prison before a break, and unrepresented ones, who are less likely to be granted parole than those with attorneys, tended to be last in line. Thus, the story of the irrational judges guided by their appetite is a case of a correlation being all too quickly interpreted as a causation, without carefully analyzing other factors. Even so, experts are humans and can indeed make judgment errors for a variety of reasons. For nearly a century, criminologists have reported that many poor people, people of color, and other vulnerable groups have been unfairly jailed.[6]

Here is where AI is called upon. Warning about the failures of human judgment, commercial companies urge shifting one's trust in humans to trust in software. The justification appears quite clear. In a first step, the algorithms could assist in improving judges' decisions. After all, we are told that an algorithm is impartial, has no prejudices, and would never alter its evaluations on an empty stomach. In the next step, with more computing power, black boxes could populate the benches and automatically output guilty/not guilty decisions in a blink, with the years of sentence attached. In the final step, this program could extend to other challenges such as litigation, solve the problems of severe case backloads, and send millions of judges, defense lawyers, and prosecution lawyers into early retirement.

Black box justice would finally resolve an old and enduring problem that even legal professionals have complained about: a surfeit of lawyers. A raft of bad lawyer jokes testifies to this sentiment:

Question: *Why didn't the shark eat the lawyer?* Answer: *Professional courtesy.*

Question: *What are five lawyers at the bottom of the sea?* Answer: *A good start.*[7]

Jokes aside, we are already moving toward black box justice. In the United States, the police arrest more than ten million people every year. After arrest, a judge first needs to decide whether the defendant is released until trial or retained in custody. Increasingly, judges and police rely on black box risk assessment tools to predict whether a defendant is likely to fail to appear at a court hearing, to reoffend at some point in the future, or to commit a violent crime. The COMPAS algorithm alone has been used in US courts for over one million defendants to inform judges on bail recommendations and prison sentencing.[8] It is difficult to know how many sentences have actually been influenced, but it is my guess that not every judge would dare to contradict the precise risk scores delivered by an algorithm. Is that trust in a black box warranted? Can recidivism algorithms actually predict better than experienced judges?

The stable-world principle indicates that the answer is no. Judges typically face decisions under considerable uncertainty. In these situations, complex algorithms are unlikely to succeed, and simple algorithms may be as good while also being transparent. But that is a hypothesis. Let us look at the evidence.

Ordinary People Can Be Just as Accurate

The COMPAS algorithm uses up to 137 features of a defendant and their past criminal record to predict whether the defendant will commit a misdemeanor or felony within the next two years. How accurate is the algorithm? To find out, one could compare its predictions with those made by legal professionals, if such a study existed. Or one could use a lower bar, comparing the algorithm with ordinary people who have absolutely no experience in sentencing. One such study does exist.[9] The researchers behind it recruited 400 people via Amazon Mechanical Turk, an online crowdsourcing service. On this platform, anyone can earn a little cash by participating in scientific studies. Each participant was given short profiles of fifty real defendants (containing only seven features such as sex and age from the more than 130 features that COMPAS uses) and was asked to predict whether each defendant would reoffend within two years. We can assume that few of the participants were highly motivated to spend much time on each case: for making all fifty predictions, they were paid as little as a dollar, with a small bonus if their predictions were accurate. As a consequence, most of these low-paid workers no doubt rushed through the task as hastily as possible. The results bowled over both legal scholars and the study's authors.[10] COMPAS had predicted the behavior of 65 percent of the defendants correctly and erred on 35 percent (counting both false alarms and misses); the ordinary people, who knew next to nothing about recidivism, got it right just about as often.[11] And, if you take the majority vote of twenty ordinary people, they predicted 67 percent correctly. The result is not limited to COMPAS alone. A review of it and eight other risk assessment algorithms concluded that the accuracy of all nine of them has been found wanting.[12]

Let's stop a second and consider what that means. Say that you are standing trial. A black box algorithm advises the judge that you are at high risk of committing another offense. Will your judge listen? The chances appear to be high. Now imagine that your judge is instead told that the greater part of twenty random people on Amazon Mechanical Turk believe that you are at high risk. What judge would trust judgments made in a few seconds by unqualified people online? Although the expensive risk assessment tool fared no better than the general public, the nature of its being a black box imbues it with a mystical aura of power. It also impedes authorities from questioning how accurate such an algorithm actually is.

Transparent Justice

Black box algorithms have set off an emotional debate about being unfair to certain groups such as the poor or people of color. Yet there is an even more fundamental issue: transparency. Without transparency, it is hard to determine in the first place whether an algorithm is fair. For instance, *ProPublica* tried to reconstruct the COMPAS algorithm and concluded it is racially biased, while other researchers arrived at the opposite conclusion.[13] Lack of transparency also conflicts with many people's understanding of justice and dignity. Most of these problems could be avoided by using a transparent algorithm.

Decision lists are one type of transparent risk assessment tool that already exists. The CORELS algorithm is a machine-learning tool that generates such lists from data on previous cases with crystal-clear logic.[14] Take the case of predicting whether a defendant will be arrested within two years. The decision list is this: If the defendant is between eighteen and twenty years old and male, then predict arrest. If the defendant is between twenty-one and twenty-three and has two to three prior offenses (whatever sex), then predict arrest. If not, then check whether the defendant has more than three prior offenses. If that's the case, predict arrest. In all other cases, predict no arrest (figure 7.1).

Note that only age, sex, and prior offenses enter the decision list. There is nothing mystical about it, no crystal ball hidden in a black box. What the machine-learning tool does is extract the most important features and determine the exact rule. Despite its simplicity, the decision list with only three features predicts arrest as accurately as COMPAS with its up to 137 features. This finding that black box algorithms for predicting future arrest are no more accurate than transparent and simple algorithms is the rule rather than the exception.[15]

The decision list illustrates what I mean by *transparency*. The algorithm is not kept secret *and* it is also understandable:

IF	age between 18–20 and sex is male	THEN predict arrest (within 2 years)
ELSE IF	age between 21–23 and 2–3 prior offenses	THEN predict arrest
ELSE IF	more than three priors	THEN predict arrest
ELSE	predict no arrest	

Figure 7.1
A transparent algorithm for predicting whether a defendant will be arrested within the next two years (created by CORELS). The combination of these four rules is called a decision list.

Feature	Risk Points for "Yes"
Pending charge at the time of the arrest?	1
Prior conviction?	1
Prior failure to appear older than two years?	1
Prior failure to appear in the past two years?	once: 2 more: 4

Figure 7.2
A transparent risk assessment tool known as the Public Safety Assessment (PSA). Shown are the four features used to predict the risk of failure to appear at a court data. If the answer to any of the four questions is yes, then a number of risk points is assigned.

> *An algorithm is transparent to a group of users if they can understand, memorize, teach, and execute it.*

Looking at the decision list, you know exactly how the prediction is made. Decision lists increase transparency, make it easier to test for potential discrimination, save the costs of paying for secret algorithms, and demystify the procedure. In the present case, the simple list is as good or bad as the complex secret algorithm. But a judge who likes to use algorithms as decision aids can at least readily use and understand these lists.

Another transparent tool that is more widely known is the Public Safety Assessment (PSA). Its purpose is to help judges decide whether a defendant should be released before trial. For instance, to predict the risk that a defendant will fail to appear at a court date, it uses only four features (figure 7.2). For the first three features, a "yes" results in one risk point each; the last feature, "prior failure to appear in the past two years," leads to two risk points;

and if the defendant failed to appear twice or even more often within that time frame, this makes four points. Defendants and judges can easily see what the features are and how these are weighted, and look up on the internet how the final risk score is calculated.[16] Using different sets of features, the PSA also predicts new criminal activity, such as a violent offense prior to the defendant's case. Like the decision list (but unlike COMPAS), it is not commercial.

The logic of the PSA is similar to that of the pay-as-you-drive algorithm (figure 4.6): a point system with a small number of features scored by simple numbers. Just as in the case of telematic car insurance, transparent risk assessment allows defendants to adjust their behavior, such as to avoid missing a court hearing. Changing their behavior for the better is impossible if the algorithm is secret. Business secrets are one reason to prevent customers from understanding an algorithm, but not the only one. Another is complexity. Even when an algorithm is made public, it can be so complex that laypeople and professionals can barely figure out how a decision is made or a score is calculated. Transparent algorithms are not confined to decision lists or point systems; other possibilities are presented throughout this book.[17]

Transparency has many virtues. In emergency situations, it is important that professionals have easy-to-memorize triage rules in order to execute these quickly and effectively. Transparency also helps determine whether an algorithm contains bias, such as racism. We can see that neither PSA nor the decision list above includes race as a feature. Still, we can't entirely rule out the possibility that the other features they look at are correlated with race. But again, their transparency makes it easier to check whether this might be the case, such as if one of the four questions was whether you live above 125th Street in Manhattan. When there are more than a hundred features, by contrast, race can be correlated with many of them, making it a huge task to identify the hidden biases.

Transparency alone, however, cannot guarantee that the resulting scores are any more accurate than those of black box algorithms; in both cases, the actual outcomes are fundamentally uncertain. In the case of the PSA, studies mostly report that it is moderate to good in its predictions.[18] A more important question is whether a risk assessment tool actually improves judges' decisions that would be made on their own, without any algorithms, and how it compares with other tools. When I searched for answers, the most striking result was the lack of studies that posed these questions.[19]

Predictive Policing

Predictive policing has been marketed as a magic bullet. *Time* magazine named it one of the fifty best inventions of the year 2011.[20] Big data companies promise that their algorithms can predict the location of future crime scenes and identify persons at high risk of becoming an offender. That would make policing more objective and efficient. Police officers would no longer need to walk the streets to observe firsthand what's going on but could conveniently stay at their office desk and see what the program says. Chicago and Los Angeles were among the first to buy in and announce their predictive policing as prestige projects.[21]

In 2012, the Chicago Police Department introduced predictive policing software to determine which citizens are likely perpetrators or victims of crimes. The software scored people by analyzing myriad factors, such as their criminal histories. The motto was "We know who they are."[22] Within a few years, about 400,000 individuals ended up on a list, their risk scores included. This list did not undergo an independent check, nor was its impact measured. Eight years later, the Chicago Police Department quietly buried the program after researchers from the RAND Corporation published a report with a devastating conclusion: there was no evidence that the program had reduced violence. Instead, the obscure inner workings of the software created a high level of public fear and mistrust.

A year earlier, in 2011, the Los Angeles Police Department had launched Operation LASER to target violent repeat offenders and gang members with "laser-like precision." The image was of police officers functioning like trained surgeons who use laser technology to remove tumors.[23] Once again, people were put on a list, and it somehow happened that 89 percent were not white, indicating that the software reflected the racial biases among white Americans. Eight years after LASER was introduced, the LAPD issued its own devastating report on the unreliability of the "laser-like" risk scores and improperly trained personnel. It shut the program down.

Nevertheless, predictive technology companies have been aggressively moving into the rest of the world, claiming to make policing objective, transparent, and effective. Not everyone buys into these promises. In

California, the Oakland Police Department canceled the program, worrying about discrimination against certain neighborhoods and eroding community trust.[24] In Germany, the Hamburg Police Department refused to purchase predictive policing software on the basis of marketing claims but did their own homework and studied its potential. The police department concluded that predictive policing cannot fulfill the hyped expectations and decided against acquiring the software.[25]

Why does predictive policing fail to meet the promises made? A typical answer is that more data are needed. In fact, just as much information is available for predicting crime as for predicting recidivism or the flu. The answer instead lies in the uncertainty surrounding human criminal behavior. Too many factors determine who commits a crime and where, and it is not possible to identify these factors on the basis of past cases and predict future behavior. The data may be "dirty" and contaminated by racial or other prejudices, which leads to faulty risk scores and a negative impact on individuals. The positive lesson in the cases of Chicago and Los Angeles is that journalists, civil society organizations, and, finally, public commissions worked together to abolish a commercial pseudoscientific project. In the cases of Oakland and Hamburg, it was even stopped before it started.

An alternative to complex black box algorithms in such situations would be psychological AI. Consider geographic profiling and the case of a serial offender who has committed six armed robberies in a city within the last weeks. Where should the police begin searching for the criminal? Psychological AI analyzes the intuitions of experts and turns their heuristics into algorithms. For instance, experts know that most offenders live in the area where the crimes happen. This intuition translates into the *circle heuristic*, which predicts that the offender lives within a circle whose diameter is defined by the distance between the two farthest crime locations. Taking that strategy, the police might start searching at the center of the circle and work from there. When researchers tested the circle heuristic, they found that it predicted as well as or even better than commercial black box algorithms, particularly when the number of offenses was smaller than ten, which is usually the case.[26] In addition, when police officers were subsequently trained to use the circle heuristic systematically, they outperformed complex algorithms at locating offenders.[27]

As with predicting recidivism, psychological AI in the form of simple rules such as the circle heuristic can provide transparent alternatives to overcomplicated and opaque geographic profiling algorithms. In my research, I have found that simple heuristics often match or outperform complex algorithms in making predictions about health care outcomes, financial outcomes, and other uncertain situations.[28] These heuristics can be used to augment expert decisions while enabling experts to understand their logic.

Why Algorithms Can Perpetuate Discrimination

Even its most enthusiastic fans largely concede that AI has a bias problem. Discrimination by gender or race has been reported for AI systems used by police, courts, employers, credit scoring agencies, and beyond. Personalized algorithms have offered the better-paid jobs to white males, and, as mentioned before, Google's image classification system notoriously identified a dark-skinned couple as "gorillas." AI is supposed to be neutral, objective, and data driven, so how can it be unfair to women, people of color, or other marginalized groups?

It is important to understand what discrimination is and what it is not. Consider the Vienna Philharmonic, one of the best orchestras in the world. A music aficionado can identify the orchestra from the very first chord by its lustrous sound. Yet someone who is tone deaf might also do so with the help of their eyes rather than ears, by spotting the small number of female musicians. It took until 1997 and enormous public pressure before the first woman, a harpist, became a full member (she had already played in the orchestra for decades at low pay and was retired soon after). The fact that more men than women have been hired by orchestras worldwide is, by itself, not proof of discrimination. It could be that among the best available players, more are male. However, if men are found to be preferred over women *who play equally well or better*, that is proof of discrimination. When blind auditions behind curtains were introduced with jurors who could not see the candidate's gender, it became clear to everyone that orchestras had indeed discriminated against women. By 2020, the proportion of women in world-class symphony orchestras had reached about 40 to 45 percent, up from 5 to 10 percent in the 1970s.[29]

Like human jurors, algorithms may discriminate against women, people of color, or other marginalized groups. This can be relatively easily found out if the algorithm is transparent. Consider the pay-as-you-drive algorithm (figure 4.6). Neither gender nor race is part of the features it considers; thus, there is no evidence that it discriminates. The absence of gender or race information corresponds to a blind audition where this information is not available. However, if there are other features that correlate with gender or race, such as income or neighborhood, bias might enter through the back door—although that can also be easily checked as long as the algorithm is transparent. If, by contrast, algorithms are secret and use many features, as COMPAS does, it can be quite difficult to detect. Discrimination is one strong reason why all sensitive algorithms should be made transparent.

Algorithms that are nontransparent by design, such as deep artificial neural networks, pose a greater problem. Here, the reason for discrimination is not that gender or race is used as a feature. The programmer does not even determine the features; the neural net finds its own. Instead, the data may be a source of discrimination. Consider philharmonic orchestras again. Assume a tech company trains a deep neural network to find the best players. It is fed the profiles of 100,000 applicants to top orchestras worldwide from the past fifty years, including information about whether they were hired. The network will soon find out that being male is a strong predictor, and thus reproduce the bias of the past.

This phenomenon occurs in other fields where women have been rare. For instance, Amazon's machine-learning specialists built an algorithm that rates applicants for software development positions and other technical jobs based on their profiles.[30] The idea was to give the machine 100 profiles, and it would spit out the top five candidates. To their surprise, the machine did not "like" women. Once again, the bias was located in the data, which contained the profiles of applicants from the last ten years, the far majority of those who were hired being male. Removing applicants' first names did not help much; the AI found ways around this, such as inferring gender from the names of all-women colleges.

Discrimination also exists in facial recognition systems that are trained to tell whether a face is male or female. These systems are used to identify perpetrators from security video footage; errors can lead to wrong accusations of crime. In a study, three commercial gender classification systems from Microsoft, IBM, and Face++ were shown portrait photos of males and females with lighter or darker skin.[31] When the systems classified a face as "male," they made only 0 to 1 percent errors if the men had lighter skin. But if the men had darker skin, the error increased to 1 to 12 percent, depending on the system. When the systems classified a face as "female," they were incorrect in 2 to 7 percent of the cases if the woman had lighter skin, but in 21 to 35 percent if she had darker skin. Every system made more errors with female faces than with male faces, and with faces with darker skin than with lighter skin.

Where does the bias come from? It can be found in the pictures used to train the systems. About half of the photos were of white males, the rest mostly white females, while individuals with darker skin, particularly females, were few.

After the study was published, the three corporations whose commercial systems were tested quickly updated their systems and reduced the bias. Even after revision, however, the IBM system still had an error rate of 17 percent for females with darker skin. IBM solved its discrimination problem in an ingenious way. It did not count all errors but only those where the system was more than 99 percent confident that it was right, which allowed the company to report an error rate of only 3.5 percent.[32] Most interestingly, the study did not appear to have an impact on corporations that were not named in it, such as Amazon and Kairos, which continued to have large error rates for darker females, confusing them with males. Naming may be shaming, but only for the named.

Neural Networks Can Increase Bias

Biased data are at the heart of discrimination, but deep neural networks may intensify the problem. Consider a neural network that was fed tens of thousands of pictures of human activities and taught to identify the activities and the genders of those portrayed.[33] These pictures had a typical gender bias, showing more men involved in outdoor activities such as driving and shooting, and more women microwaving and shopping. When the network then had to identify gender and activities in a new large set of pictures, the bias increased. For instance, when the activity was cooking, 67 percent of the pictures were of women. The network, however, concluded that 84 percent were women, misidentifying about half of the male cooks.

Why would a network intensify discrimination? One reason is that researchers evaluate a network's performance by the number of correct answers, not by the degree to which it discriminates. The network can actually increase performance by increasing discrimination. Say a network knows nothing except that two-thirds of the cooks are female. To attain the best results, it would guess that every cook is female, which means that two-thirds of the answers would be correct. This of course increases the bias to a maximum. To avoid amplifying the bias, the network could randomly guess that the cook is a woman for two-thirds of the pictures and that it is a man for one-third. In that case, however, it would get only about 56 percent of the answers right.[34] In general, if there is a bias in the data, amplifying it can lead to better performance than by trying to be "fair."

AI is not alone in its bias problem. The people at tech corporations who want to transform every aspect of our lives are predominantly male. *Wired* magazine reported that at leading machine-learning conferences, only 12 percent of the contributors were women, and that among Google's machine-learning researchers, only 10 percent were women.[35] This is a move backward. Recall that in the early 1980s, almost 40 percent of graduates in computer science were female. Dr. Timnit Gebru, a staff research scientist at Google and one of the authors of the gender classification study, was one of the 1.6 percent of Google's workforce who are women of color. A new study that she coauthored found that Google's large language models—which try to produce what looks like meaningful new text and conversation—run the risk of reproducing the racist and sexist language embodied in the huge amounts of text from the internet on which they are trained.[36] Moreover, this training consumes enormous amounts of computing power and, thus, electricity, thereby producing substantial rises in carbon dioxide emissions. All of this business tends to benefit wealthy organizations, whereas climate change hits poor communities first. When Google's leadership saw the paper, it decided to censor it and fired Gebru. Thousands of Google employees and supporters from academic and civil organizations signed a letter of protest in which they did not mince words: "Dr. Gebru is one of the few people exerting pressure from the inside against the unethical and undemocratic incursion of powerful and biased technologies into our daily lives."[37]

Against Uninformed Consent

In health care, *informed consent* refers to an ideal of how doctors and patients interact. Patients are informed in an understandable way about the treatment options, including their benefits and harms, and then permit the doctor to go ahead with their preferred option. Informed consent involves more than signing a form or clicking "I accept." It is *shared decision-making*. The second half of the twentieth century has seen the spread of this ideal, overthrowing earlier paternalism where doctors decided and patients had to agree—if they were even told what treatments would be performed on them.[38] Informed consent goes hand in hand with human dignity and an educated citizenry.

The informed consent forms in hospitals are a special case of terms-and-conditions contracts, also known as terms-of-service contracts. Similar to when you lease a car, you get a contract that defines each side's benefits and obligations. To enable informed consent, the language needs to be understandable, clear, and brief. The Harding Center for Risk Literacy, which I direct, is one of the organizations dedicated to making information about risks understandable and evidence-based, and the information communicated by health authorities and institutions is indeed improving from decade to decade.[39]

However, the twenty-first century has seen powerful attempts to reverse this positive development in online contracts such as terms-and-service agreements and privacy policies. Have you ever tried to read what you agree to online? We are constantly reminded to read the terms of service, but that is easier said than done. Consider *sign-in-wrap* agreements, which websites require you to accept. Unlike *clickwrap* contracts, where users have to click "I agree," sign-in-wraps simply say that by signing in, the user agrees to the contract. Even if readers took the time to read these contracts, which are legally binding, could they understand what they are signing?

Study Privacy Policies for Thirty Days a Year

In health care, a rule of thumb is that consent forms should be written at the level of eighth grade at most. An analysis of the 500 most frequently visited US websites in 2018, such as Amazon, Airbnb, and Uber, showed that the average sign-in-wrap was written at the level of "fifteenth grade," which corresponds to the language in academic journals. In 70 percent of these contracts, the *average* sentence length was longer than twenty-five words.[40] The authors of the study estimated that virtually all contracts, more precisely 498 out of 500, were unlikely to be understood by consumers. These contracts appear to be deliberately written in a way that dissuades the few who try to read them from continuing. In addition, privacy policies—which are required by law if users' data is collected—are much too long and abundant. To read the average privacy policy may take about ten minutes. To read all privacy polices one encounters in a year would require on average about thirty full workdays, according to the estimates of two Carnegie Mellon professors.[41] What's more, on some websites, you may be offered to opt out of targeted ads, yet some services have been reported to override your intention, that is, to simply ignore your "Do Not Track" setting and continue tracking you without your knowledge.[42]

When we log in online, we enter the age of uninformed consent. At the same time, courts holds us responsible, whether or not we have read the terms and conditions. That is, the law expects consumers to read contracts, but suppliers have no obligation to make these comprehensible.

In health care, patients can say no to a treatment; on the internet, users may be excluded from services if they do not accept the terms. If you do not agree to your smart bed sharing minute-by-minute data—your movements, position, heart rate, noises, and other audio signals—with third parties whose identity is not revealed, then you are told that the company cannot guarantee your safety or provide desired features and services to you. You may ask why you can't own a smart bed, thermostat, fridge, or TV that does not send personal information to unidentified third parties.

To endow users with informed consent would be detrimental to many sources of profit in the digital world. That by itself explains why tech companies have little incentive to state in clear terms what they are doing. And it is not only tech companies: almost everyone in business on the internet is cashing in on personal information.

The One-Pager

Legislation has been slow to protect citizens from feeling helplessly adrift in this ever-changing business fashioned to be opaque, deceptive, and confusing. This hesitancy is surprising given that proper regulation can work. Consider the practice of overly lengthy and opaque privacy policies. In 1999, it took only about two minutes to read Google's privacy policy. By 2018, reading time increased to thirty minutes. Yet that year, the European Union's General Data Protection Regulation went into effect, and Google reacted with a new policy that not only took half the time to read but was also more concise and comprehensible.

During my term on the advisory council at the German Ministry of Justice and Consumer Protection, we went even further and proposed making the one-pager the rule.[43] The terms of service should be written on a single page, maximum 500 words (to avoid small print), and the language used must be understandable to the average person. The terms need to lay out what personal information is extracted and to which third parties it is sent, as well as who owns your data, pictures, and videos. So far, a few companies have followed suit and produced one-pagers. But more legal courage is needed to end the age of uninformed consent. To enforce transparent one-pagers, contracts that do not stand up to these requirements should be considered invalid by courts, and those who drafted them should be fined.

Why Do We Use Black Box Algorithms When We Don't Need To?

On the advisory council, we discussed many critical questions. Should consumers have the right to know why they are denied a loan or not invited for a job interview? Should credit scorers be obliged to reveal the features and weights of their algorithms to the public rather than only to data protection agents? Should business secrets be valued above the rights of citizens to understand their scores? Although such ethical issues were at the forefront, I soon noticed that one issue was taken for granted: that complex black box algorithms are accurate. The idea that transparent rules might do just as well did not even remotely occur to most governmental officials.

Faith in Complexity and Opacity

In 2018, the analytics company FICO, Google, and various universities organized a prestigious competition, the Explainable Machine Learning Challenge. With the help of FICO, which provided data from thousands of individuals, including their credit history, the task was to create a complicated black box model for predicting loan default and to explain the black box.[44] Competitions between algorithms are nothing new, but this was one of the first landmark events that acknowledged the need to make sense of complex black box models. Even so, underlying the Challenge was the assumption that predicting loan default actually requires complex models. Only one of the teams that entered the contest, a group of researchers from Duke University, took a radically different approach. They developed and tested software that is understandable and includes visualizations that allow people to play with credit factors to see how these influence loan application decisions. Not only was the AI they developed transparent, but it was just as accurate at predicting loan default as deep neural networks and other complex black box models. The team won the FICO Recognition Award for "going above and beyond expectations with a fully transparent global model and a user-friendly dashboard."[45]

The Challenge illustrates a deep faith in two propositions:

Faith in complexity: Complex problems always need complex solutions.

Faith in opacity: The most accurate algorithms must be inherently incomprehensible.

Put these two faiths together, and you get the belief in the *accuracy–transparency dilemma*: the more accurate the algorithm, the less transparent it will be. This dilemma is widely—and wrongly—believed to be generally true.[46] It holds in well-defined games and other stable situations, but not under uncertain conditions. As the one team participating in the Challenge showed, when predicting loan default, understandable AI can be as accurate as black box AI. Here is the general insight:

Transparency-meets-accuracy principle: Under uncertainty, transparent algorithms are often as accurate as black box algorithms.[47]

Nevertheless, black box algorithms populate courts and other locations where high-stakes decisions are made because of unconditional faith in complexity and in opacity. A second reason is that firms exploit this faith. Their salespeople knock on the doors of courts and police departments, just as pharmaceutical sales representatives persuade doctors to recommend their products to patients. Organizations that develop freely available and transparent risk assessment products usually do not have the same amount of resources to pay for extensive promotion. For instance, the decision list for predicting recidivism shown in figure 7.1 is free and little known, while the software license of COMPAS is expensive and widely purchased. There is also another reason for favoring black boxes: defensive decision-making. As a judge, if you are inclined to grant a defendant bail but the risk calculator establishes high risk, then you may change your mind to be on the safe side: if the defendant does commit another crime or threaten a witness, you would otherwise have some explaining to do.

"Less Is More" Is the Key to Explainable AI

Faith in complexity and opacity remains deeply engrained and has misled leading scientists, bank and other business executives, the military, and many others. In the United States, for instance, the Defense Advanced Research Projects Agency (DARPA) project on explainable AI (XAI) assumes that understandability comes at the cost of accuracy.[48] As a consequence of this misunderstanding, it still features complex black box algorithms and tries to explain their workings in simple ways that are not necessarily correct. For instance, an explanation of how a deep neural net predicts loan default or classifies an object as a tank may have little to do with how the neural net actually works. These explanations are rough guesses at best, and incorrect at worst.

The true alternative is to replace complex algorithms with transparent ones whenever appropriate. This solution is based on the empirical finding that simple, transparent rules can match or beat black box algorithms at making predictions under uncertainty. I've often seen this happen.[49] If you work for a retailer, you know how important it is to predict which previous customers are likely to return in a given time frame. Many retailers have databases containing tens of thousands of customers, making it expensive to send all of them flyers or catalogs, especially when some are not likely to purchase anything again. A study with thirty-five retailers found that a simple rule used by experienced managers, known as the *hiatus heuristic*, predicted customers' new purchases even better than complex machine-learning methods and sophisticated marketing models.[50] The hiatus heuristic is a version of the recency heuristic described above and relies on a single powerful cue, the most recent purchase: *if a customer has made a purchase within nine months, the customer is likely to purchase in the future, otherwise not.* (The hiatus—the number of months—can vary from business to business.) This rule is transparent and can be easily explained, dispensing with the need to try to explain more complicated methods in simple terms that are not necessarily correct.

Herein lies the key to a new understanding of explainable AI: to routinely test whether transparent algorithms work as well as complex ones in a particular situation, and if so, to work with the simpler tools. That is the real future of explainable AI.

Counting Keys

Transparent algorithms can be broken down into several families. The *one-good-reason* family, as the name indicates, is made up of algorithms that base their prediction on a single but powerful reason.[51] As we have seen in the case of predicting the flu, one of these, the recency heuristic, can outperform big data analytics. A second family consists of algorithms using only a few reasons that are given different but simple weights, such as the pay-as-you-drive algorithm in figure 4.6 and the PSA. A third family is made up of short decision lists, such as the one for predicting recidivism. In machine learning, decision lists have also been around for a long time.[52] In what follows, I describe another family, known as *tallying*, that simply counts the reasons for and against a potential event.[53] All of these rules correspond to psychological AI, which, like human psychology, is particularly fit to deal with predictions under uncertainty.

The Keys to the White House

On November 8, 2016, more than a few people around the world could not believe their eyes. The polls, the election markets, and big data analytics had predicted Hillary Clinton's victory by a large margin. "If you believe in Big Data analytics, it's time to begin planning for a Hillary Clinton presidency," announced columnist Jon Markman in *Forbes*.[54]

At the end, big data nose-dived. Admittedly, predicting who gets the keys to the White House is easier said than done. It is less like a lottery, where we know the probabilities of winning or losing, than like predicting the course of the flu or another viral infection. Statistician Nate Silver and his media business team FiveThirtyEight had correctly predicted the Obama victories but missed Trump's win in the primaries and, on election day, predicted a 71.4 percent probability for Clinton. Two weeks before the election, when his estimate was 85 percent in favor of Clinton, Silver discussed in detail how the probability depends on the assumptions made. If his model considered only data since 2000, as opposed to 1972, Clinton's chances would rise to 95 percent. If it assumed a normal distribution of votes (a concept introduced by the mathematician Carl Friedrich Gauss; see figure 3.1) instead of fat tails (a concept featured in Nassim Taleb's book *Black Swan*), that would get Clinton 87 percent, and if it assumed that state outcomes are not correlated with each other, that would boost her chance of winning up to 98.2 percent.[55] Silver's valuable reflections bring an important insight home that is too often forgotten: *big data does not speak by itself.* Instead, the outcome depends on the assumptions made. More data and more computing power do not guarantee the truth. That's why statistical thinking matters.

Silver's discussion also reveals something equally interesting: it focuses on reasons for or against statistical models, not on the actual reasons why people might vote against Clinton and for Trump. It dispenses with psychological, political, or economic theory. The alternative is to begin with the psychology of voters' reasons. That is precisely what Alan Lichtman, a distinguished professor of history, did.

Lichtman's was one of the rare dissenting voices among the mass of experts who predicted a clear win for Clinton. He predicted that Trump would win. It was not the first time he was right; he had predicted all elections correctly since 1984.[56] His method does not rely on number crunching with big data, nor does it deliver ostensibly precise probabilities of winning. It simply predicts who will win. The system is called Keys to the White House and is based on a historical analysis of the reasons why Americans vote the way they do.

A key is a reason that matters for voters. There are thirteen of these, each stated as a proposition that can be answered by yes or no. A yes favors election or reelection of the candidate from the incumbent party, and a no does not.

Key 1: *Incumbent-party mandate.* After the midterm elections, the incumbent party holds more seats in the US House of Representatives than it did after the previous midterm elections.

Key 2: *Nomination contest.* There is no serious contest for the incumbent-party nomination.

Key 3: *Incumbency.* The incumbent-party candidate is the sitting president.

Key 4: *Third party.* There is no significant third-party or independent campaign.

Key 5: *Short-term economy*. The economy is not in recession during the election campaign.

Key 6: *Long-term economy*. Real annual per capita economic growth during the term equals or exceeds mean growth during the two previous terms.

Key 7: *Policy change*. The incumbent administration effects major changes in national policy.

Key 8: *Social unrest*. There is no sustained social unrest during the term.

Key 9: *Scandal*. The incumbent administration is untainted by major scandal.

Key 10: *Foreign or military failure*. The incumbent administration suffers no major failure in foreign or military affairs.

Key 11: *Foreign or military success*. The incumbent administration achieves a major success in foreign or military affairs.

Key 12: *Incumbent charisma*. The incumbent-party candidate is charismatic or a national hero.

Key 13: *Challenger charisma*. The challenging-party candidate is not charismatic or a national hero.

You may note something peculiar. Almost all of the keys concern the incumbent party and its candidate; only one (the last key) concerns the challenger. Some of the keys require no judgment, such as whether the incumbent-party candidate is the sitting president, while others, including charisma, do. Lichtman dealt with this problem by defining the rare charismatic leaders such as Dwight Eisenhower and John F. Kennedy, and settling yes or no before the election. In the case of Clinton versus Trump, neither was classified as charismatic.

The question is, how should the keys be combined into a prediction? The knee-jerk reflex of many data scientists would be to develop a scoring system like those in online dating or credit scoring, where each key is given "optimal" weights. Yet there haven't been many presidential elections to base the scores on, and irregular voting behavior makes the task even more difficult. Instead, Lichtman developed a transparent algorithm that simply counts the no's:

If six or more keys are negative ("no"), the challenger will win.

This tallying rule is radically simple. In late September 2016, weeks before the election, Lichtman considered the keys to be settled and made a count.[57] Six keys turned against Hillary Clinton, the incumbent-party candidate:

Key 1: The Democrats were crushed in the midterm elections.

Key 3: The sitting president was not running.

Key 4: There was a significant third-party campaign by libertarian Gary Johnson, anticipated to acquire 5 percent or more of the votes.

Key 7: There was no major policy change in Obama's second term.

Key 11: Obama did not have any smashing foreign policy successes.

Key 12: Hillary Clinton is not charismatic in comparison to, say, Franklin D. Roosevelt.

Six negative cues meant that Trump was predicted to win. Six is the minimum required, which reflects that the outcome of the election was close and certainly not easy to predict. It is worth pointing out a caveat. The rule is intended to predict who will win the majority vote, which Trump did not. But no prediction system is perfect, and Lichtman's tallying rule came closer to the end result than the polls, prediction markets, or big data analytics did.[58]

The Keys to the White House is transparent. Its transparency enables us to see the theory behind the predictions. And there is indeed a theory behind them, unlike in the typical machine-learning exercise, which is about getting the best prediction, no matter how. As mentioned, almost all of the keys concern the incumbent party and its candidate. They relate to the economy, social unrest, foreign policy successes, scandal, and policy innovation. That means if people perceived that the country fared well in the previous term, the candidate of the incumbent party will be elected. If a challenger like Trump wins, the victory has little to do with him personally but solely with how people perceived the incumbent party's performance in the previous term and with their expectations regarding the party's candidate.

Many were flabbergasted on election day, asking what in the world had driven Americans to vote for a man who had insulted women, Muslims, and the Pope, among many others. The radical logic underlying the tallying rule suggests that this is the wrong question. US voters did not vote for Trump; they voted against Obama and Clinton. Forget TV debates, raising large amounts of money, and investing in advertising. Do not believe that campaign managers and consultants had a large impact on the outcome. If the theory underlying the keys is right, there is a positive message to political parties: focus on governance, not on costly advertising and campaign tactics.

Wanted: The Right for Transparency in Scoring

In a black box society, people in power use software to better predict and modify the behavior of others, without disclosing their algorithms. Why are particular people denied bail or credit and others not? Why does YouTube's recommendation system steer us toward less fact-based and more extreme videos?[59] The characteristic feature of black box societies is not the absence of transparency but its asymmetry, like a one-way mirror.[60] Black box societies have existed since time immemorial. For centuries, ordinary Europeans were unable to read. For them, the Bible and other sources were black boxes written in Latin or Greek, which only the educated from wealthier circles understood. With the help of the printing press, invented by Johannes Gutenberg, translators such as Martin Luther opened the black box. Thanks to this technological breakthrough, books and translations could gradually be made accessible to everyone. Now people could find out what the Bible actually said. Gutenberg's invention leveled the disparity between priest and layperson, between initiated and follower. When the internet spread in the 1990s, physicians envisioned a similar revolution: "The internet can help us to level the disparity between doctor and the patient, the infallible and the uninformed."[61] We may need another Gutenberg to open the black boxes that predict and modify people's behavior.

One of our proposals at the Advisory Council for Consumer Affairs was that *all black boxes used to score people and have serious consequences for their lives should be made transparent to the general public and subject to quality control.*[62] Serious consequences can occur in health scoring, credit scoring, recidivism risk assessment, and predictive policing, among others. The proposal concerns algorithms that are imposed on people, not those chosen deliberately for entertainment or personal growth, such as video games or love algorithms. Disclosure should include all features of an algorithm as well as its logic (such as a decision list or a point system, as in figures 7.1 and 7.2).

When we discussed the proposal with international scoring companies, they raised legitimate objections. One was that their algorithms are business secrets. But that was exactly the point of the proposal: to change the law and rank people's rights above commercial profit. Moreover, telematic and health insurers already make their scoring algorithms fully transparent, and it does not seem to hurt them. The next objection was that people would not understand the source code (the program of the algorithm), so making it transparent would keep it in the black box. But we were not asking them to disclose the source code, which would indeed be of little help for most consumers, but rather to disclose the features and the logic of the algorithm. Knowing the features would allow consumers to find out why, for instance, they pay high interest for a loan. It might simply be because they live in an apartment house in which some tenants didn't pay back their loan in time. That's called geo scoring.

The companies' next objection was that there is no way to make deep neural networks transparent. Although that is true, we found that commercial scoring systems for credit, health, telematics, and recidivism rarely, if ever, rely on neural networks. A final objection was that if users knew the features, they could change their behavior and game the algorithm. Yet that is exactly one of the stated purposes of credit scoring, health scoring, or telematic scoring: to lead people toward healthier or more financially sound behavior. For instance, if people find out that owning too many credit cards is one reason for their low credit score, they can act on this. Gaming is only a problem when algorithms use proxies instead of the real features, such as when a health insurer gives customers bonus points for joining a gym rather than for actually exercising there.

In the European Union, the GDPR marks the start of a new era in privacy law. It requires that "meaningful information about the logic involved" in automated decision-making be provided and that people "have the right not to be subject to a decision based solely on automated processing" that affects them "legally" or "similarly significantly."[63] Yet the moment a human decision maker appears at the end of the chain, as with judges and police, these regulations no longer apply. In general, the GDPR remains highly abstract, showing that it has identified a regulation problem but not yet resolved it. Meanwhile, hundreds of mathematicians have called to stop all work on predictive policing, demanding that the algorithms undergo a public audit to prevent abuses of power.[64] People have protested against the secrecy of credit scoring companies, demanding the legal right for all of us to look at the logic behind the algorithms that predict and modify our behavior. Governmental authorities are called upon to become more aware of the motivation behind these algorithms and of the persistent lack of quality control.

Secrecy and mystique is one aspect of the black box. Yet the term has a double meaning: it also refers to a recording device. In the next chapter, we deal with this other side of the black box.

8   Sleepwalking into Surveillance

It's time to start paying for privacy, to support services we love, and to abandon those that are free, but sell us—the users and our attention—as their product.
—Ethan Zuckerman, inventor of the pop-up ad, "The Internet's Original Sin"

I imagined the future SmartFridge stationed in my kitchen, monitoring my conduct and habits, and using my tendency to drink straight from the carton or not wash my hands to evaluate the probability of my being a felon.
—Edward Snowden, *Permanent Record*

One popular episode of the British TV series *Black Mirror*, called "Nosedive," plays in a future world where each person wears smart contact lenses that zoom in on every other person in sight, be it at a party or on the underground train. The lenses immediately display the others' names and their social scores. Similar to Amazon products, everyone has a score, from 1 to 5. No matter what you do, your behavior is rated by others: in a restaurant, the waiter scores you and you score the waiter; at work, your boss scores you and you score your boss; in a taxi, the driver scores you and you score the driver. Scoring is made easy; all you need to do is type a score into your smartphone and point it at the person in question. Living in this future, Lacy, an attractive young woman, wants to move to a better part of the city. Her score is 4.2, too low for an affordable apartment in that neighborhood, for which she needs at least 4.5. Lacy dutifully practices smiling in her mirror. She does everything to please other people and behave in a way that will win her more points. Increasingly, Lacy's life revolves around her score. She has become remote-controlled by others' ratings in order to move to a location that others have rated as desirable. In the episode, as its title indicates, her efforts did not end well.

Many who have seen "Nosedive" are appalled by this vision of a world in which only one thing counts: one's score. But the same people who react so aversely engage in various forms of scoring themselves without much thought. They rate the Uber driver, restaurants, doctors, hospitals, hotels, and tax accountants and distribute likes to posts and hearts to pictures. Scores are becoming the hard currency for trustworthiness, as they were for Lacy. Eyeglass computers that record what wearers are seeing and hearing already exist and, similar to security cameras, alter how people comport themselves in public so that socially accepted behavior increases.[1] They could easily be adapted to display the names and scores of everyone around. Wouldn't it be handy to see immediately how trustworthy a person is, instead of finding out the hard way?

Social Credit Systems

Credit scores such as the FICO score try to measure your creditworthiness, that is, the likelihood that you will pay your debts in time. The engineer Bill Fair and mathematician Earl Isaac introduced these scores in the 1950s when they created Fair, Isaac and Company—thus the name FICO. The score is a number between 300 and 850, determined by payment history, debt burden, length of credit history, and recent credit inquiries, among others. Your neighbors' credit scores also matter; if the person next door defaulted on a credit, your score may go down. Almost everyone has a score, even those who are not aware of it. Banks, telephone companies, and employers use your score to determine whether you will be considered worthy of a loan, a phone contract, a job, or an apartment. When you order a pair of shoes from an online shop, you may not notice that your credit score is being checked while you make a few clicks to finalize the purchase. Online shoppers are subject to these credit checks all around the world. The result may determine the payment options you are given. You won't even see the options you are denied.

Social credit scores try to measure your trustworthiness in all respects, not just financial honesty. They are like FICO scores but integrate whatever data can be gathered about you. These may include previous speeding tickets, your criminal record, your commitment at work, your engagement in voluntary social services, the degree to which you fulfill your family duties, your political statements on social media, the websites you watch, and your entire digital footprint. Crossing the street at a red light means losing points; visiting your elderly parents brings points. Playing video games for too long may cost you points; watching parenting lessons or learning a foreign language will add points. If you have online friends with low scores, that may also lower your own score. This system influences your behavior best if your social credit score is made public, like in "Nosedive," and if everyone can see the count of likes or hearts on social media. People with a low score may feel ashamed and may worry about what others will think of them. The moment you begin to worry, you have swallowed the bait.

"Nosedive" is science fiction, but social credit scores are not. Led by China, various governments, including Thailand, Myanmar, Vietnam, Venezuela, and Tanzania, have introduced social credit systems for individuals and businesses or announced intentions to do so.[2] In addition to public shaming, incentives and penalties are added. Those with high scores are rewarded perks, such as access to cheaper credits and free health care checks, while thousands of people with low scores have been denied access to airplanes and fast bullet trains, and their children denied access to the best schools. People have begun to unfriend others on their social media on the basis of their scores. In lonely hearts ads, young men and women advertise their social credit score next to their age, weight, and interests. Nobody is forced to do so, but an ad without a score may arouse suspicion. In an online survey of participants of social credit systems in China, 94 percent said that they have changed their behavior to positively influence their social credit score—they adhered to traffic regulations, reduced time playing online games, volunteered for community service, shared different content online, and used mobile payment apps, among others. Fifty percent said they had already shared their score with family and friends.[3] What counts as proper behavior is determined by governments and big technology companies, unlike in the world of "Nosedive," where anyone can up or down someone else's score.

Chinese society is rooted in a Confucian version of collectivism, where the greater good dominates individual rights. It has a long history of social control, best known from Mao's cultural revolution. Digital technology for mass surveillance took off with the Beijing Olympics in 2008, for fear of terrorist attacks on participants and the public, and became amplified in response to the 2009 Xinjiang conflict with separatist groups. China is also striving to control the problems emerging from large parts of the population relocating from rural to urban areas. With digital technology, surveillance is now 24-7. Its goal is to improve people's moral behavior, eliminate corruption, and create a culture of "sincerity" and "harmony." Surveys indicate that the far majority of Chinese citizens are in favor of this system, even more those with higher education.[4] Many believe that it provides a true alternative to democracy, fostering sincerity, harmony, and economic growth for all.

Like credit scoring, social credit systems have a double function: to protect people and influence their behavior. Credit scorers offer their services to protect businesses from fraudulent customers who don't pay their bills; otherwise, honest customers would have to pay higher prices to compensate for fraud. As a consequence, more people pay their bills on time, avoid owning too many credit cards, and do not overspend. Similarly, governments justify social credit systems because these protect citizens from selfish free riders, criminals, and terrorists. To avoid punishment, such as being barred from purchasing a plane ticket or seeing one's face and address on a downtown billboard for having run a red light, more citizens respect laws and social norms, which in turn increases social stability. In the Western press, these systems are often called Orwellian projects, which misses the point. Social credit systems are more in the spirit of B. F. Skinner's *Beyond Freedom and Dignity*. Skinner, a Harvard psychologist, declared freedom to be our greatest problem: it allows people to exploit others and their environment. His proposed solution was a system of strict behavior control that rewards fair and responsible social behavior. If Skinner had lived to see social credit systems, he might have been delighted to see his methods of behavior modification applied to more than a billion people worldwide. The difference is that, in Skinner's view, positive reinforcement alone, not punishment, is effective in encouraging social behavior. Apart from that, the digital age offers the technology to turn his vision into reality.

The Slippery Slope into Social Scoring

Social credit scores with carrots and sticks attached often make the headlines in Western countries, largely because they are seen as incompatible with the Western ideal of a democracy that values freedom and privacy. Yet Western attitudes toward privacy appear to be shifting closer to Chinese ones, adapting to what digital technology offers. And it seems to be happening faster than we are willing to recognize. Take Germany, the land of data protection and privacy. Germans learned the hard way about the value of privacy during the Third Reich and the German Democratic Republic, regimes that surveilled, controlled, and suppressed their citizens. In these eras, shrugging one's shoulders and saying, "I have nothing to hide," would have meant conforming to the injustices of either system. As a reaction to this history, the first article in the German Constitution addresses human dignity.

Given their past, one might think that Germans are united in rejecting surveillance, and particularly a social credit system for behavior modification. Yet they appear to be shifting their stance. In 2018, only 9 percent of Germans thought that a social credit system was a good idea for their future.[5] By 2019, the figure had already risen to 20 percent.[6] Similar enthusiasm was recorded in Austria, particularly among the political right.[7] Even Germans appear to be getting used to the fact that their smartphones, smart TVs, and smart cars are already recording their steps—why not get an additional few bonuses from the government for behaving properly?

Credit companies, health insurers, car insurers, Payback, Google, Amazon, Microsoft, Facebook, IBM, Apple, and everyone who asks you to accept cookies collect data about you. These companies build their own secret profiles of you. This may seem like no big deal because the data are not added up into a single credit score. Now imagine a data broker who collects data from thousands of these sources and creates a profile for every person, which can be used to calculate a social credit score. This is no fantasy; the companies already exist. Data brokers such as Acxiom and Oracle Data Cloud advertise these services and are working on constructing a comprehensive profile for every citizen. In the United States, Acxiom has gathered health data on 250 million consumers, including prescriptions, medical history, and data from hospitals, labs, and health insurers.[8] It also includes criminal records, voter records, item-level purchase data across hundreds of thousands of stores and pharmacies, and real-time location data from mobile apps and from tens of thousands of sensors physically placed in malls, airports, movie theaters, and college campuses. Even homebodies are not free from surveillance: Acxiom collects second-by-second TV viewership data and everything else your smart TV records about you. Personal profiles are sold to banks, insurers, credit card issuers, health care providers, and governments, including your email address, phone number, IP address, and postal address. Acxiom claims to have collected data from more than 700 million people worldwide and up to 3,000 data points for each individual. Not all of what they collect is accurate. Nevertheless, Facebook has purchased data about their users from brokers, containing all these inaccuracies.

In popular culture, Facebook has been branded the villain in profiling its users, yet data brokers track us much more intrusively. We need to talk about the entire data broker industry, not just Facebook and the apps it owns, such as Messenger, WhatsApp, and Instagram. And we are silently profiled by other organizations we would never think of. Graduates from major US universities have discovered to their surprise that their alma mater collects dossiers about them, more than one hundred pages in length and compiling everything from marriage and divorce to financial status broken down in liquid and illiquid assets, to find out how to target them to donate.[9]

Companies such as Acxiom use the same technology that enables China's social credit system. Yet they collect the data as surreptitiously as possible, while China is at least upfront about doing so. Openly or secretly, commercial companies and governments alike build a world of surveillance. Being watched will be our future—unless people, legislators, and courts intervene to stop the scoring of the public.

Privacy Isn't Worth a Damn

Privacy refers to the right not to be subjected to invasions of one's private sphere by government, corporations, and individuals. It includes the right to be let alone, to prevent intrusions to one's physical space, and to maintain a sphere of intimacy. Privacy is not a central value in every culture, nor has it been one throughout Western history. It is primarily a product of the nineteenth and twentieth centuries. When postal services introduced the postcard in the 1870s, the reaction was moral outrage because people believed the motivation was to spy on their private correspondence. In the 1980s, hundreds of thousands of angry Germans took to the streets in protest against a governmental census that asked for private data, such as date of birth, gender, personal status, and education. Americans have been taught to guard their Social Security number carefully; otherwise, someone might open a bank account or apply for a loan in their name.

Most Americans react with horror and repugnance when hearing about the erosion of privacy by social credit systems in other countries. The media disparagingly note how useful these systems would have been for the Spanish Inquisition, the KGB, and the East German secret police. As often occurs with such outbursts of aversion, however, there is a paradox. The same people who complain about social credit systems are willing to hand over their personal data to commercial companies without a blink: what they buy, where they are all day and with whom, what websites they visit, whether they pay their bills on time, and when they visit a doctor and for what purpose. The discrepancy between privacy declarations and actual behavior has a name:

Privacy paradox: The same person who claims to be worried about privacy is not willing to pay a cent for it. Rather, they give their private information away on social media and other platforms without much thought.

The term *privacy paradox* was originally used in a study that reported that US undergraduate students said they wanted to keep personal information private but nevertheless posted it on Facebook.[10] The students did not seem to realize that Facebook is a public space where parents and future employers can read the entries. Yet the privacy paradox is more general; it even exists when people are aware that they are being surveilled and their data collected. More people in Germany than in any other European country are aware that free internet services take their data and analyze it.[11] More than three-quarters say they are concerned about privacy in the digital age.[12] If they are concerned about privacy in the social media and are also well-informed about what happens with their data, then they should be willing to pay a fee for the service, as they do for Netflix, radio, TV, and other services, instead of paying with their data. How much is privacy worth?

In 2019, I conducted a survey with an insurance company, where we asked a representative sample of 3,200 people in Germany: What do you think is the greatest danger of digitalization? Half of the respondents (51 percent) said loss of privacy, namely that their personal data are collected and made accessible to companies and governments.[13] Then we asked: If you could pay for social media—Facebook, WhatsApp, Instagram, and all others—in exchange for keeping your personal data, how much would you be willing to pay per month?

Here is the sobering truth:

Nothing	75 Percent
Up to 5 euros	18 Percent
6 to 10 euros	5 Percent
More than 10 euros	2 Percent

I was taken aback by the answers. Three-quarters were not willing to pay a penny! If privacy is truly the greatest concern for half of these people, why is it not worth anything? After all, the social platform companies need some source of revenue. One might speculate that young people are probably those least willing to pay. That was not the case. The unwillingness to pay with a fee instead of with one's personal data was similar for all age groups, from eighteen-year-olds to those over sixty-five. Only 7 percent were willing to pay six euros or more. At first I thought that the privacy paradox might be specific to Germans, for whatever reason. But it is not.

In another study by Norton LifeLock, participants from sixteen countries or regions were asked whether they are concerned about their privacy. On average, 83 percent said they are somewhat or strongly concerned. The lowest concern was in the Netherlands, with 66 percent, but that's still a majority. The highest was in China, Mexico, and Taiwan, where over 90 percent were concerned. Then the same people were asked whether they would be willing to pay at least a dollar a month to ensure protection of their personal information on all social media. On average, less than a third were willing to pay a single dollar. For instance, in the United States, 85 percent of participants said they were concerned about privacy but only 28 percent were willing to pay a dollar for privacy. The willingness to pay varied strongly between countries and regions (figure 8.1). People in the United Arab Emirates, Brazil, Mexico, and China were most willing to pay a dollar, compared with very few in Germany, Australia, Canada, New Zealand, and the Netherlands. All in all, the privacy paradox appears particularly pronounced in Europe, the United States, and the Commonwealth countries.

Why would so many people say they are very concerned about privacy and at the same time be reluctant to pay even the smallest amount? When asked why, a typical answer is "I have nothing to hide." But then why

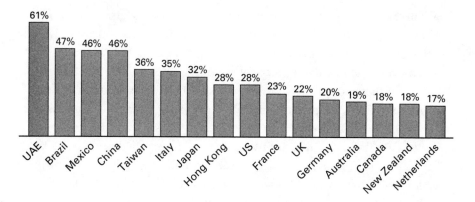

Figure 8.1.
The privacy paradox by country/region. Percentage of people aged eighteen and older who are willing to pay $1 per month for social media in return for keeping their personal data. Numbers based on Norton LifeLock, Cyber Safety Insights Report: Global Results, 2019.

would one be concerned about privacy in the first place? And how readily would any of us hand over our smartphone to a stranger on the street so that they can download its contents? That's essentially what Google, Facebook, and others do without asking us—that's the deal we make with them by paying with our data rather than a fee. You may think that you are just one of the millions of users and that what you do is of no interest to the rest of the world. Yet companies may try to infer whether you are inclined to depression and an easy target for persuasion, if you are pregnant so that you buy their pregnancy products, if you are an undecided voter who can be swayed, if you are having an affair and with whom, what your boss and colleagues think of you, and whether you have a severe medical condition. Those data can be of great interest to advertisers, health insurers, political parties, private detectives, employers, and quite a few others. Similarly, your smartphone and your smart car send minute-by-minute information about where you are, even if you aren't actively using them. Access to that information may attract other groups of people, including burglars who want to know whether you are at home or stalkers who want to know where to find you. The business of stalkers—following, watching, or harassing others, often former partners, to create fear and do harm—has become much easier. In fact, social media and surveillance apps on phones have created a stalker's paradise. As one female victim reported, "My abuser magically knows where I am."[14] These are just a few of the multitude of reasons why it's in our best interest to maintain some degree of privacy.

The two features digital technology offers, convenience and surveillance, conflict with privacy. Many people feel helpless and cannot see any other options. Others prefer immediate convenience over long-term loss of privacy. Still, it is remarkable that most people are not willing to pay even a dollar a month to regain their privacy. Some of them might not trust social media corporations to honor such a deal or believe that other agencies will collect their data anyhow. Whatever the reasons, the privacy paradox may well soon become a strange blip in history. Perhaps surveillance will one day be seen as nothing more than a natural byproduct of convenience— nothing worth worrying about.

Newspeak: Privacy Is Theft

In George Orwell's *1984*, the ultimate art of thought control is called "Newspeak." Newspeak is a new language that determines what you can think and what you cannot. Its purpose is to express the worldview of an ideology and to make thinking differently unthinkable. Newspeak does so by altering the meaning of concepts and even inverting them to stand for the contrary. In Newspeak, war is peace, freedom is slavery, and ignorance is strength. It could well be that the privacy paradox will be resolved in a similar way. Privacy is theft. Surveillance is sharing and caring.[15]

When users pay with their data as opposed to their money, withholding personal information might well be considered theft. This business model requires eroding users' privacy to make money from their data. Facebook is a prime example. It originated in an act of privacy violation.

Hot or Not was a website, founded in 2000, where users could score people's attractiveness on a scale from 1 to 10. Whistling at and catcalling women passing by is a tired old game, but now at least women could join in the rating. Soon, Hot or Not morphed into a sequence of dating services and inspired the creation of the first social media sites.[16] It also stimulated the invention of YouTube, which was originally intended as a video version of Hot or Not where people uploaded videos of themselves, but it failed at drumming up sufficient video material (if TikTok, which makes it easy to create selfie videos, had existed then, YouTube might have ended up as the largest dating platform).

Hot or Not also provided the inspiration for the first version of Facebook, conceived by Mark Zuckerberg in 2003, then an undergraduate at Harvard. It was called FaceMash and used pictures of female Harvard undergraduates. Zuckerberg had hacked into the websites of Harvard's dormitories to gather pictures from the "face books," which were paper-based directories containing photos of the women in Harvard's dormitories. Without asking the women or the houses for permission, he then posted their photos in pairs next to each other and asked users to rate which woman was "hotter." Harvard University quickly shut down the site for violating individual privacy and copyright. A year later, the first version of what was now called Facebook was launched for Harvard students, and then rapidly spread across the globe.

Since this initial breach, Facebook has attempted numerous attacks on users' privacy to monetize their data. For instance, in 2007, Facebook introduced Beacon, a program that allowed advertisers to track users through the internet and disclosed their purchases to their personal network of friends. For one, in violation of the Video Privacy Protection Act, Beacon showed the titles of the videos users had rented from Blockbuster Video on its news feed. After massive protest, Zuckerberg apologized, and Facebook paid $9.5 million as part of a settlement. Two years later, Facebook changed its users' privacy settings without warning or requesting permission, a practice that made personal posts public, which the federal government called "unfair and deceptive." In response to the protest, Zuckerberg declared that the rise of social media means that people can no longer expect privacy.[17] He may in fact be right, but note that he was talking about a select group of people. Facebook managers tend to use all means to shield their private lives, and their employees typically have to sign nondisclosure contracts that bind them to silence, unless they want to risk legal prosecution. The surveillance business model requires that users alone, not the managers and actual customers of tech companies, relinquish their privacy. That's the deal, plain and simple.

Since then, privacy violations have continued and been tolerated by many users for whom Facebook is the main platform for conducting their social lives. In 2018, the British *Observer* and the *New York Times* revealed that Cambridge Analytica, a political consultancy, had extracted personal data from millions of Facebook accounts and secretly used it to craft political ads targeted at unknowing users. Facebook had known about this problem since 2015 but did little to remedy it. Instead, the technology company threatened to sue the Guardian Media Group, which owns the *Observer*, if they publicized the scandal.[18] Only after another wave of public outcry did Facebook promise to reduce the amount of data shared with third parties and make it easier for users to control privacy settings. Following that scandal, hashtags to delete Facebook went viral. But these had next to no impact: a month later, Facebook announced that its membership had actually grown.

Facebook collects heart rate data and ovulation data, among others, from apps and tries to infer from posts whether people are feeling stressed, defeated, overwhelmed, anxious, useless, or a failure.[19] All that information is supposed to enable its customers to target users with the right ads at the perfect moment. That reminds me of earlier eras, such as in sixteenth-century France, when neighbors would have known everything about you and listened to every creak of the stairs. Under those conditions, little room for privacy existed, and the iota that was left was easily filled by gossip. In such a culture, everyone was watching and policing everyone else 24-7. At that time, people like you and me were owned and reigned by sovereigns, the sole inhabitants who were not constantly subject to public scrutiny (at least they were walled away from the masses). These sovereigns dictated the terms of ordinary people's lives in exchange for protection and other services. Today, tech companies are the new nobility that dictate the terms, written at such length and in barely comprehensible language so that the modern subordinate simply clicks and accepts. These terms may include the rights to your personal data, photos, videos, and other products. But there is an alternative to this autocratic vision of the internet.

Pay with Your Money, Not with Your Privacy

For some, the dawn of the digital age felt like the French Revolution of 1789, without the blood. For others, it felt like 1989, when the people of East Germany toppled a regime that had controlled the media and kept their citizens under surveillance. Twenty years later, during a wave of demonstrations in the Middle East and North Africa known as the Arab Spring from 2010 to 2012, social media began indeed to play a substantial role in coordinating public protests. The Arab Spring was the proof that fueled a new excitement about the liberating force of the internet. Everyone can know everything. Secrets are a thing of the past. Open access to knowledge for all humans is possible. The world will be like science in its ideal form, not politics and marketing, free of censorship and profit interests—or even more liberated than science, where a handful of publishing companies charge prohibitive fees for open access to articles. The internet once again embodied the big dream of liberty, equality, and fraternity.

Sergey Brin and Larry Page, the young founders of Google, had a similar vision when they were in their mid-twenties. In 1998, as mentioned previously, they criticized other search engines that relied on advertisement as income, arguing that advertising money distorts what is listed at the top. In their own words, "advertising funded search engines will be inherently biased towards the advertisers and away from the needs of the consumers" and that it is "crucial to have a competitive search engine that is transparent and in the academic realm."[20] After the dot-com crash hit Silicon Valley in 2000, startups were forced to shut, traffic to Google's website was surging, and its top venture capitalists pressured Google's founders.[21] Finally, Brin and Page caved in, did an about-face, and built an advertising-based business model whose algorithms are now among the best-kept secrets in the world. Borrowing from Christian terminology, one might call this turn to an ad-based business model the "original sin" of the internet.[22] Users were driven out of the paradise of knowledge to became the object of surveillance. After Page and Brin made personalized advertisement their business model, a spot in paradise was reserved for Google's new customers, the companies that paid them for placing targeted ads.

Surveillance Capitalism

To deliver a service in exchange for personal data rather than cash is a new form of economy, known as *surveillance capitalism*.[23] It was perfected by Google and imitated by Facebook and other companies. Data about users are collected and profiles are constructed that allow advertisers to target those who are most likely to click on the ad and, eventually, make a purchase. The advertiser pays the platform for every click, impression, or purchase a user makes. To function, this business model requires secretly invading people's privacy, and more invasion promises better personalized targeting. Unlike industrial capitalism, which manufactures and trades physical goods, surveillance capitalism collects and analyzes personal data. It is extremely important to understand that surveillance capitalism is not an inevitable consequence of a smart world. It is a new business model invented by human beings to make profit. A smart world enables unforeseen possibilities for surveillance but would function more healthily without it.

How did we get there? Consider the icon of surveillance capitalism, "cookies," the bits of code that pass information between your computer and servers of tech companies. So-called third-party cookies are used by companies to track and compile a user's long-term record of browsing, such as which websites the person has looked up and when. When cookies were developed in the 1990s, there was serious concern about privacy. For instance, in 2000, the US government banned cookies from all federal websites, and in April 2001, Congress was debating bills that included regulating cookies.[24]

Yet something unexpected happened. With the trauma of 9/11, everything changed. Fear of terrorism pushed privacy concerns back in favor of security. Surveillance technology suddenly became of great interest for governments. The US government launched the Patriot Act, limiting laws that protected civil liberties. The United Kingdom, France, and Germany followed suit, implementing similar acts. The Total Information Awareness (TIA) program and subsequent programs allowed US tech companies to collect data with few restrictions, and in turn obliged them to deliver the data to the government if requested. The program's official goal was to detect and identify foreign terrorists before they attempt an assault. After the public learned of its existence, the TIA program was abruptly canceled. Fear of terrorism instigated an unprecedented collaboration between Google and the US intelligence community, particularly the National Security Agency (NSA). Google later collaborated with the Obama presidential campaign to identify wavering voters who could be persuaded to vote for Obama in the 2008 and 2012 elections, which further cemented the link between the tech industry and government.[25] Revolving doors tightened the link once more. By 2016, twenty-two former White House officials had left the administration to work for Google, and thirty-one Google executives had joined the White House or federal advisory boards. Another twenty-five former intelligence or Pentagon officials moved to Google, and three Google executives took up positions at the Department of Defense.[26] The iconic nightmare of 9/11 enabled and boosted the rise of surveillance capitalism.

Fear of terrorism also led to quite a few citizens' willingness to accept government and commercial surveillance above privacy and freedom. Surveillance cameras began to populate buildings and streets. When we hear about governmental surveillance, most in the West automatically think of China. But the United States has about fifteen CCTV surveillance cameras installed per 100 people, which is slightly more than in China, and European countries such as the United Kingdom and Germany have put up about half as many, with numbers growing.[27] When it was finally revealed that the NSA secretly collected phone records of millions of Americans and their personal data from Facebook, Google, and other top tech companies, government officials justified this intrusion into citizens' lives as a means to protect them from terrorists.

There is a deep irony to this. There is little evidence that mass surveillance technology has prevented more terrorist attacks than traditional antiterrorist expertise previously did. For instance, the claim by the US government that fifty-four terrorist plots had been thwarted thanks to surveillance turned out to be fake news.[28] An in-depth analysis of 225 individuals recruited by al-Qaeda and other terrorist groups who were charged in the United States with having committed an act of terrorism revealed that virtually all of them had been caught by traditional investigative methods, such as tips from local communities. Mass surveillance appears to have played a role in at most 1 to 2 percent of all cases.[29] Similarly, as we saw a chapter earlier, the predictive policing algorithms developed since then have not lived up to their promises. These results are consistent with the stable-world principle: combating terrorism is not a well-defined problem that you can easily solve with big data. Despite lack of evidence that mass surveillance works, governments around the world, joined by private companies, nonetheless seized the opportunity to do so, politely called "bulk collection."

This brief account of how we got into both governmental and commercial surveillance makes one thing clear: neither form of surveillance needed to happen.[30]

Pay a Fee for Service

It's not only privacy that is at stake in surveillance capitalism. It's also your time. A tech company can generate more money from advertisers if more people spend more time on its site. Therefore, its engineers enlist a bag of tricks to keep users, young and old, on their sites as long as possible (see next chapter). The result is that you waste hours of your life in a state of perpetual distraction and disruption by advertisement and notifications. As one study jokingly reported, the attention span of the average user is rapidly declining and already on average shorter than that of a goldfish.[31] The increasing number of people who report having difficulties concentrating for more than a few minutes is a dangerous development—think of teachers and judges not being able to focus their attention, not to speak of surgeons and pilots. For instance, 78 percent of 439 US perfusionists, who operate heart-lung machines during open-heart surgery, said that using cell phones for texting, surfing, posting, and the like while working is a safety risk for the patient. At the same time, one-third of them also reported having witnessed other perfusionists get distracted by using cell phones during bypass surgery.[32] Moreover, the psychological tricks used to glue users to the phone and compete for likes appear to have disturbing psychological consequences for some teenagers. Since smartphones became available in 2007, depressive symptoms, self-harm, and seriously considered suicide have increased and feelings of liking oneself and self-competence have decreased. Yet these observations are correlations, not necessarily causes. Whatever the reasons, digital natives appear to be more anxious than previous generations and less comfortable with face-to-face conversations.[33]

What can we do about all of this? One popular proposal known as *data dignity* calls for social media companies to pay users for their data.[34] That amounts to a minimal solution that would reduce neither surveillance nor the loss of privacy, time, and attention span. On the other hand, it would at least introduce some degree of fairness. How much compensation would this proposal realistically earn each user? I have met people who believe that their data are worth at least $100 a year, and that's what Facebook should pay them. Let's do a rough calculation to check.[35] Facebook has about 2.7 billion active monthly users.[36] In 2019, its revenue was about $70 billion, and profit (net income) was $18.48 billion.[37] If Facebook were to pass on half of the profit to its users, how much would everyone get? The answer is $9.24 billion divided by 2.7 billion users, which is about $3.40 per year, or 28 cents per month. In other words, users would get about a cent a day, making them the worst-paid workers on the globe. This is a very rough calculation, but even if you refine it, there won't be much more in the bank. In other words, getting paid for our data promises to be a lousy deal.

What is the alternative? In my opinion, we need to make a radical move. To stop the surveillance business model, tech companies need to adopt the business model of charging a fee for their services. This would be the key to rescuing privacy and preventing a future of commercial surveillance that can easily flow over into governmental surveillance. Consider an analogy with the postal system. If the post office changed its business model to surveillance capitalism, you would no longer have to pay to send your mail; delivery would be free. In exchange, to earn money, the postal system would openly or secretly read all of your mail and sell the content to interested third parties. To preserve privacy, you have to pay with money instead of personal data. In exchange, social media platforms would need to provide verifiable assurance that our usage is not tracked. This solution would take care of the entire chain of damage, from surveillance to loss of time and attention to psychological problems that hurt our children.

To get there requires strict legislation that curbs surveillance capitalism and citizens who are willing to pay a small fee. How much would that be? For a very rough estimate, we can make another back-of-the-envelope calculation. Facebook earns most of its money from advertising. Let's be generous and reimburse Facebook for its entire revenue: $70 billion divided by 2.7 billion users would make about $26 a year, or about $2 a month. If one takes account of the $47 billion in expenses Facebook had in 2019 and the fact that part of these expenses would vanish if there were no need for collecting, storing, and analyzing users' personal data anymore, then Facebook would make a larger profit or the fee could be reduced. Although some users might quit, most would likely stay attached to the platform. This rough calculation indicates that a fee-for-service model could be a viable alternative.[38]

A flat fee is not the only alternative. Micropayments can also eliminate the intrusive and privacy-compromising ads that rule the internet. Mozilla and other companies have announced a $100 million grant program to bring the dream of an internet without ads to fruition. This would also resolve the ongoing arms race between ad-blocking software and countermeasures by advertisers, such as ads dressed up as news.[39]

To enact this change, we need governments that are willing to resist social media lobbyists along with their customers and aggressively regulate corporate privacy practices and business models. This extends beyond social media to all companies that try to get their hands on our data.

Acquiring the right to say no to cookies is not enough; quite a few companies then deliberately make it a time-consuming chore, annoying users by asking them to go through a series of clicks and scrolls at every step until many eventually give up and say yes with a single click. At the same time, we ourselves need to care about the increasing degree of surveillance and its psychological consequences, and we need to understand that paying with data is at the heart of the problem. Taking that step will be equally challenging, given that most of us are now accustomed to free services and are reluctant to pay for privacy. It's time to wake up and make this move. We can enjoy social media without such negative consequences.

Mass marketing of our attention began with the business model of Google and was intensified by many competing social media platforms. Pay-for-service would stop that. Yet that measure by itself would not stop a related surveillance business that has moved from online into our offline world. Our behavior is now monitored by almost every product prefixed with the term *smart*. Have you ever wondered why your new mattress or TV came with a privacy policy?

Sleepwalking into Surveillance

Sleepwalking, or somnambulism, is a mixture of sleep and wakefulness. Sleepwalkers walk with their eyes open. When they wake up far away from their bed, they tend to be surprised and disoriented: How did they get there? Sleepwalking into surveillance means walking away from freedom and privacy with one's eyes open and, when waking up, being bewildered about how one got there. For some, it began in their playroom.

Smart Dolls

Mattel's Barbie was based on the comic-strip character Lilli in the German tabloid *Bild*, a sassy and exhibitionist blonde bombshell, designed to satisfy the taste of *Bild*'s adult male readers. Like Lilli, the first version of Barbie had unrealistically long, thin legs and a chiseled waist; the 1963 version even came with a book titled *How to Lose Weight*, recommending "don't eat."[40] Barbie contributed to girls' feeling that their own bodies were inferior and too fat.[41] In 1992 a version of Barbie appeared that could speak phrases such as "Math is hard. Let's go shopping."[42] Those words of wisdom reinforced girls' feelings of not being up to boys in math and relegated them to consumers. In 2015, Mattel introduced another version to the US market, named Hello Barbie but soon to be dubbed "eavesdropping Barbie." This interactive Barbie can respond to the worries, hopes, and feelings that children entrust to their doll. Unknown to them, however, the doll records their intimate conversations and sends them over the Web to a server for analysis. These conversations are sold to third parties, including parents, who can choose to receive the audio recordings daily or weekly.[43] Hello Barbie got the Big Brother Award for enabling parents to spy on their children.[44] One can imagine how a child might feel after eventually finding out that the beloved doll—and parents—eavesdropped and betrayed them.

But perhaps something more fundamental will change in a developing mind. Children may not even feel betrayed because they take constant monitoring for granted and thus develop no concept of private space and time. In this possible future, children adapt their feelings to the potential of technology. Play is surveillance. Surveillance is security.

Eavesdropping is one intrusion; changing the nature of a child's conversation is another. Children have always talked to their dolls and stuffed animals, attributed to them a personality, built a close relationship, and engaged in role-play and conversation. Now the algorithms in Hello Barbie drive the conversation and talk to children about products they want to market: pop culture, other Barbie products, the latest movies, and musical artists. Without being able to freely practice their own imagination and creativity, children are likely to lose whatever skills they have developed so far. They have to adapt to the limited possibilities of the technology and to the values of the corporation behind it: fashion, the desire to be thin, and an obsolete view of women's purpose in life.

Smart Homes

In one session of the German Ministry of Law and Consumer Protection, we had a closed meeting with two men clad in hoodies, whose CVs sported an intriguing combination of professions: both were hackers as well as professors of internet technology. They came to testify on questions of internet security and what the law can do about it. At the end of the meeting, I asked them what they thought was the greatest danger in the commercial digital world. Their answer came without a blink: smart homes.

A smart home contains devices and appliances such as refrigerators and coffee machines that are networked and linked to the internet. Thanks to this technology, you can switch on the coffee machine with your smartphone before getting out of bed, ask your digital assistant what time it is, or access the internet with your TV. When you are away on vacation, smart alarm systems allow you to check day and night whether some suspicious-looking person is hanging around the house. Being connected to the internet also allows your TV or fridge to transmit data minute by minute about your actions to its home company, such as Samsung. In the near future, you might well get targeted ads on the surface of your fridge or your walls.

Smart homes have been advertised as security systems, so the hackers' answer was surprising. Why did they believe these pose such a great safety risk? Their concern was that smart homes invite security disasters. Even apparently benign smart devices, such as smart light bulbs, electricity meters, and alarm systems, can provide entry points for hackers. If you have programmed your light bulb to turn on and your alarm system to switch off when you are at home, then a hacker who can access your light bulb and set it "at home" can disable the alarm system. Smart light bulbs, fridges, TVs, security cameras, and other appliances have little protection from hackers; otherwise they would be prohibitively expensive. Because they are mass-manufactured, a hacker only needs to buy a single device and figure out its weaknesses, and can then apply that knowledge against all those who own that product.[45]

In a study, fifty experts in security, cybersecurity, antiterrorism, and the social sciences worked together and developed twenty-four scenarios of security breaches enabled by smart home technology.[46] The scenario judged to be most likely is smart home blackmailing. Hackers can use ransomware to tap into smart home applications, including webcams to record and blackmail the victims. For instance, researchers from the University of Michigan hacked into Samsung's Smart Things platform, allowing them to control the light bulbs and program their own PIN code in the front door lock to open it.[47] The researchers concluded that there are too many vulnerabilities in smart homes to be able to fix all of them readily. Another scenario judged as highly likely is sex crimes, such as a prospective sexual assaulter gathering information from smart home appliances about the intended victim's habits and whereabouts in order to know when to strike. That scenario also includes hacking into appliances such as webcams or smart TVs to watch and record the inhabitants' sexual activities. In a final scenario, which made it into the novel *Blackout*, smart electricity meters are hacked to cause a power outage.[48] As a result, food supplies grow scarce, people fight over gas, riots start, smartphones run out of energy, and nuclear plants run on emergency electricity until they begin to melt down and radioactive clouds emerge. *Blackout* is science fiction, but smart homes with smart electricity meters are reality. The Fukushima Daiichi disaster has shown what a breakdown of emergency cooling systems in nuclear power plants means.

To provide their services, smart homes have to surveil their owners. Smart home technology allows a degree of surveillance unprecedented in history, which is attractive because it is often useful, from parents monitoring the breath of their sleeping infant on their smartphone to later tracing the whereabouts of their teenagers. But there is a slippery slope from useful monitoring to less desirable forms. The same kinds of apps that allow parents to oversee where their teenage children are at all times can be adapted to apps that allow people to check where their ex-partners are right now. When everything becomes connected, the world becomes much more convenient and much more vulnerable.

Yet few smart home owners are aware that, together with convenience, they bought surveillance. In a survey I conducted, only one out of every seven people was aware that a smart TV records their conversations. It is not so hard to find out. "Please be aware that if your spoken words include personal or other sensitive information, that information will be among the data captured and transmitted to a third party," Samsung's privacy policy reads.[49] And if you are a person of interest for secret services, they can observe you using your smart TV even when it is switched off, as former CIA director Michael Hayden confirmed: "You want us to have the ability to actually turn on that listening device inside the TV to learn that person's intentions. This is a wonderful capability."[50]

Yet the smart home was not initially conceived for spying on its inhabitants. In 1991, the US computer scientist Mark Weiser dreamed of the smart home, part of an entire smart world, where tiny computers are in everything from clothing to kitchen devices.[51] All these would be built into the environment as invisible servants. In his dream of "ubiquitous computing," the internet would weave itself into the fabric of everyday life, being freed from specific devices such as personal computers and smartphones.

People, in turn, would be freed from an unhealthy life sitting all day in front of computer screens in windowless offices. Weiser took into account the possibility that overzealous government officials and marketing firms could make unpleasant use of this technology, but he believed that cryptographic techniques would safeguard private information. Weiser had a clear view into the future of smart technology but could not foresee either its exploitation by surveillance capitalism, invented a decade after his dream, or its alliance with governments. Such surveillance also comes in other, more or less subtle forms.

Big Nudging

Social credit systems influence people using carrots and sticks: if rewards do not work, undesirable behavior is punished. Yet there is a subtler way to influence people. *Nudging* is a system of control that needs neither carrots nor sticks but instead takes advantage of people's psychology to steer them into some desired behavior.[52] Big nudging is the combination of big data (or digital technology in general) with nudging. The idea is to identify people's weak points and exploit these to influence behavior on a large scale.

Assume that the top management of a tech company is worried about a political candidate who has announced her intention to regulate the company and make it pay taxes like everyone else. The management would of course prefer that people elect another candidate who opposes regulation. But they can't use carrots and sticks, nor can they buy all the votes. What else could they do? Big nudging is the answer: to find something in people's behavior that could be used to make them want to vote for the management's favorite candidate themselves. Here is an example:

> *Urge to click*: Many people lack click restraint, that is, they cannot resist the urge to click immediately. As a consequence, about 90 percent of all clicks are on the first page of a search result, and half of these on the first two entries.[53]

The next question is, how to exploit the urge to click? The company can modify the page rank algorithm of its search machine so that the positive reports about their favored candidate more likely appear on the first page, and the negative reports on the pages further down—and vice versa for the candidate they want to eliminate. The far majority of users, who rarely get past the first page, will consequently more likely land on the positive reports of the favored candidate and not read the negative ones. This will shift their opinion, which in turn will make them more likely to vote for that candidate. There is no need to persuade, bribe, or coerce a single voter.

But will it work? It won't work with voters who have already made up their minds, but it might well influence those who are undecided. With many elections won by small margins, these are the ones who can make the difference.

Psychologist Robert Epstein (who, as you recall, went on a date with a woman who had used another woman's picture in her dating profile; he also happened to be Skinner's last graduate student) put this idea to the test.[54] He showed that search engine manipulation worked in lab experiments. But can it also work in real life?

Elections in India are the largest democratic elections in the world, with more than 800 million potential voters. In 2014, there were three candidates for prime minister: Narendra Modi, Arvid Kejriwal, and Rahul Gandhi. Modi is best known as a Hindi nationalist, Kejriwal for his campaign against corruption, and Gandhi as the uniter of the country, reducing caste and religious tensions. Epstein recruited 2,150 undecided voters throughout India right before the election. He seated them in front of computers where they could search for information about the candidates using an internet search engine. The web pages had been manipulated: for one group of voters, more positive texts about Modi popped up on the first page and more negative ones on later pages. For the other groups, the same was done for the other candidates.

The manipulation went virtually unnoticed; 99.5 percent of participants showed no awareness of it. Sadly, it worked. Epstein reported that the biased search engines could shift the voting preferences of undecided voters by about 20 percent. How much is that in absolute terms? Think of 100 voters, ten of whom are undecided. Without search engine manipulation, half of these ten would eventually vote for your favorite candidate.[55] After the manipulation, this number would increase from five to six, that is, an absolute increase of one percentage point. If more voters are undecided, the effect will be larger. One percentage point is not much, but it could be enough to flip a close election.

Given this possibility, it is hard to believe that the executives behind search engines would not be tempted to use their power in critical elections.

After all, it would be difficult to detect when it happens because the page ranking algorithms are top secret. In most countries, where the majority of people rely on the same search engine, a single company can tip the ballots.

Big Nudging, Small Effects

How can people be influenced to vote in the first place? To use big nudging, one again needs to identify a psychological bias and exploit it. Let's take the 2010 US congressional elections. On the day of the election, sixty-one million Facebook users were sent a message encouraging them to vote, including a link to the local polling places. The reminder had no effect at all.[56] Only when users were also shown up to six pictures of close Facebook friends who said they had already voted was turnout increased by 0.39 percentage points. Again, that is not much, but for these millions of users, the authors estimate that the social message increased turnout by about 60,000 people. And if these in turn motivate their best friends, that snowball effect may add up to many more. Once again, a psychological predisposition was exploited:

Imitate your peers: Do what your closest friends do.

In fact, showing the familiar faces of friends who said they had voted made all the difference. This study nicely illustrates how understanding psychology is necessary for technology to work successfully. That can be done for the greater good of everyone, such as increasing participation in political decision-making. But big nudging can also be used by someone who wants to influence the outcome of elections toward a particular candidate. If Facebook prompted only users who are likely to vote for a specific candidate or party, they could increase voter turnout in favor of that party.

This study led to much concern about Facebook's and other tech companies' allegedly unprecedented power to influence and manufacture votes.[57] At the same time, the study documented once again that big nudging, like campaign advertisements, tends to have only little effect on manipulating voters both for or against political candidates.[58] Yet in close elections in winner-take-all systems, such as in the US presidential elections of 2016 and 2020, even small effects can change the winner. When small effects are scaled up by big numbers, they gain force.

Can Facebook Manipulate Your Emotions?

You may have read about a study run by researchers at Facebook and Cornell University demonstrating that Facebook has the ability to make you feel good or bad simply by tweaking what shows up in your News Feed.[59] Over 689,000 Facebook users unknowingly participated in this mass-scale "emotional contagion" experiment. For some of them, the researchers filtered out a number of posts in their News Feed containing positive emotions; for others, they omitted posts containing negative emotions. Then they read the users' subsequent posts and counted the number of words with positive and negative emotional content. Users who saw fewer positive words subsequently posted messages with fewer positive words, and those with reduced negative words posted messages with fewer negative words. The *Guardian* and many other reputable newspapers reported how Facebook has discovered the means for making us feel happier or sadder, and both awe and horror of this power over our emotions fueled a heated debate.[60] Many were angry that Facebook had manipulated their News Feed and treated them like lab rats, unaware that they had in fact consented to such experiments when signing Facebook's Data Use Policy. After all, it is in Facebook's interest to find ways of manipulating their users' emotions in order to allow advertisers to target users when they are in the right mood: some products are more likely to be sold to those who are sad and vulnerable or to those in a state of bliss.

What the debate mostly overlooked was an important question: How big is the effect in the first place? When positive posts were reduced in the News Feed, the percentage of positive words in people's status updates decreased by 0.1 percentage points, from an average of about 5.2 percent to 5.1 percent positive words, and the percentage of negative words increased even less, by only 0.04 points. That is all the study actually showed. In addition, it is unclear whether that minuscule change is really due to emotional contagion, as claimed. If Facebook filters out the positive post I receive from Joe, this by itself can lead me to not respond with any positive words to Joe. And if the same happens to all other users, this adds up to fewer positive words across the board.

What do these experiments tell us about Cambridge Analytica's ability to influence elections? They suggest that the effect was smaller than Alexander Nix, its former CEO, boasted, such as for the campaigns of US

Senator Ted Cruz, who paid millions to Cambridge Analytica to influence voters, and of Trump, whom Russian hackers allegedly helped to win in 2016. Despite the promises made by companies advertising the power of their predictive analytics and persuasive advertising to sway voters, there is a general reason to doubt that these effects are big. The stable-world principle suggests that in an uncertain world where voters are subject to many kinds of influences, predictive analytics will be of limited success, and subsequent attempts at voter manipulation will have only a small effect. The experimental evidence reported above confirms this theoretical statement. However, it remains worrying because even small returns of big nudging can provide the tipping point in close elections.

My conclusion is that big nudging can indeed manipulate the public without being noticed, but its effect has also been oversold by tech companies to make political parties and others buy their services. The more powerful instruments to control people's behavior are the sticks and carrots of social credit systems, combined with 24-7 surveillance and justified by governments to protect their citizens. The hidden attempts to manipulate our behavior and feelings by big nudging and the overt attempts to do so by social credit systems once more illustrate the importance of staying in control of technology.

Into the Future

In the early days of the internet, a *New Yorker* magazine cartoon showed a dog sitting at a computer with the caption "On the Internet, nobody knows you're a dog." At the time, that was a joke. Twenty years later, Edward Snowden sat at a computer while working for an NSA contractor and was able to read a stranger's emails, listen to his phone conversations, and watch him sitting in front of his computer, a toddler on his lap.[61] The spy software is known as XKeyscore, which allows almost unlimited surveillance of anyone anywhere. Remember the photo Facebook's CEO Zuckerberg posted to celebrate Instagram's growth to 500 million users? In the background it showed his laptop with its camera taped over. The times when dogs were free to go undercover on the internet are long gone.

In *Beyond Freedom and Dignity*, Skinner argued that freedom is not the solution but the problem. People are free to exploit their employees, destroy the environment, and wage wars. Instead of revering freedom, we should

Figure 8.2
An artistic rendering of the evolution from *Homo sapiens* to *Homo digitalis*, sleepwalking into digital surveillance. © iStock by Getty Images.

reward fair, environmentally friendly, and peaceful behavior so that harmful behavior is not even learned. Skinner truly believed that strict control of behavior is the way to change the world into a better one. As we will see in the next chapter, social media companies have adopted his techniques but with a different motivation. Skinner's techniques are employed to keep people on the app and to create an urge to return quickly. Protecting the environment plays little role. On the contrary, the internet and the vast servers and data centers that support it account for about 3 to 4 percent of global greenhouse emissions, similar to the amount produced by the airline industry, contributing to climbing temperatures, increased wind speeds, and more wildfires.

Where does the invasion of privacy lead us? We can start with two peculiar features that may shape the future. First, there is the flawed ideology of *technological solutionism* that every social problem is a bug that needs to be fixed by technology. For instance, in response to the problems of fake news and hate speech, tech leaders have proposed building more AI tools.[62] Second, tech companies and governments work much more closely together than most of us think. Take these two features together, and we get a glimpse at what might happen (figure 8.2).

When I interviewed experts on social scoring on behalf of the German Ministry of Law and Consumer Protection, a possible scenario emerged—while keeping in mind that all predictions about the future are notoriously uncertain. The scenario consists of three phases.

Phase One: China Is Overt about Surveillance, the West Is Secretive

Let's begin with the world's largest data projects run jointly by governments and tech companies, the social credit system in China and Aadhaar in India. Aadhaar builds a database of India's 1.3 billion citizens, including demographic and biometric information, such as iris scans.[63] Both programs collect data in order to protect their citizens and control their behavior. In a first phase, which is happening at present, China, India, and other governments, helped by their tech companies, are developing and experimenting with the software and hardware, with remarkable public acceptance. The awareness that one is scrutinized by the "digital eye" changes behavior, reduces violence, and boosts moral behavior and economic output. The successful tracking of infected people during the coronavirus pandemic in China demonstrated the benefits of close surveillance and gave the government an opportunity to scale and speed up the technology. Trust in government and the educating effect of surveillance is high. For instance, in an anonymous online survey, 80 percent of Chinese approved of social credit systems, while only 1 percent disapproved (the remainder were indifferent).[64]

In the first phase, most people in the West pay little attention to what happens and consider this development a "Chinese problem." Few seem to notice that similar developments are happening in the West, despite Edward Snowden's revelations that the US government was secretly working with Apple, Facebook, Google, Microsoft, and other tech companies to run national and global surveillance programs. Surveillance does not mean that someone personally reads your messages or listens in on your phone calls, although that may occur. It means that your metadata are recorded, that is, minute-by-minute data of which websites you looked at, where you are, where you spent your night, with whom, and for how long—your entire digital footprint. When Snowden first revealed the unprecedented scale of government surveillance, the majority of Americans shut their eyes and instead favored criminally prosecuting him for treason.[65] In Britain, the Government Communications Headquarter (GCHQ) created a similar mass surveillance program, code-named KARMA Police, that tracks their citizens' instant messages, emails, Skype calls, phone calls, visits to porn sites, social media, chat forums, and everything else people do online.[66] GCHQ also hacked European companies' computer networks and the largest SIM card manufacturer in the world, Gemalto, by secretly stealing encryption keys that protect the privacy of cell phone communications.

All in all, the exchange of information between governments and tech companies goes in one of two ways: direct orders kept secret from the public, and upstream collection of data kept secret from the companies surveilled.[67] The PRISM program enabled the NSA to ask Google, Facebook, Microsoft, Skype, and others to hand over user data, including email, photos, and video and audio chats. Upstream collection is even more invasive and managed by NSA's Special Source Operations unit, which embedded secret wiretapping equipment in corporate facilities around the world. How does that work? Imagine you open a web browser, type in search terms, and hit enter. No longer does your request go straight to that server: On its way, it has to pass through secret servers installed at major telecommunication corporations throughout countries allied with the United States. The programs installed on these servers make a copy of the data coming through and check them for suspicious key terms. For instance, if you are curious about the spy software XKeyscore and type the term into your search engine, you are likely to be singled out, and an algorithm decides which of the agency's malware programs to use against you. The malware can access all your data, not only metadata, and is delivered to you along with the website you requested.[68] All that takes less than a second, without you noticing a thing.

These actions have been conducted in absolute secrecy. In no case has the public had a chance to voice its opinion in the process. Mass surveillance was given the harmless name "bulk collection," reminiscent of Newspeak. The Chinese government, in contrast, has been open about constructing its surveillance system. Western governments also tend to tolerate Western surveillance capitalism and may even take advantage of it by accessing the data collected by a select group of immensely rich companies that reign the internet and buy up competitors rather than creating a competitive market. Democracies have failed to prevent the concentration of too much power and data in too few hands. When Twitter closed President Trump's account, many considered it long overdue while others were alarmed and called it undue censorship. The real issue is that Twitter and similar companies virtually alone have the power—once confined to absolute monarchies—to decide who has the right to speak and who must remain silent.

During this first phase, Westerners grow accustomed to surveillance by tech corporations thanks to the pay-with-your-data business model. It's *personalized* advertisement, not advertisement per se, that buries the original dream of the internet connecting all people to expand their horizons. Instead, the algorithms nudge us into filter bubbles where we see more of the same, not different, viewpoints. If you live in Israel and are rooting for existing Israeli policies, you are likely to see videos on Instagram of rocket launches by Hamas adjacent to hospitals. If you are pro-Palestinian, you might instead see reports on an Israeli sniper who murdered Gazan children.[69] For each side, it is difficult to be aware of what they don't see. Personalized advertisement increases political polarization. And it's not only social media; virtually every organization, from academic institutions to charities to online shops, now uses cookies to collect data about visitors on the internet.

At the same time, surveillance changes behavior. The technologically sophisticated can bypass surveillance and withdraw into a dark net. But ordinary people notice that everything they do is being recorded and begin to think twice about what websites they watch and what they post and say. As Google's former CEO Eric Schmidt reminded us: "If you have something that you don't want anyone to know, maybe you shouldn't be doing it in the first place."[70] That moral advice could have come from many an authoritarian government as well.

Phase Two: Social Credit Systems Boost Autocratic Systems Worldwide

In a near-future second phase of the scenario, the social credit system has been fully implemented in China. Commercial systems such as Sesame Credit and governmental systems have morphed into a single unified system. China will export hardware and software to other governments with similar regulatory interests, which have by now realized that there is an efficient alternative to democracy, namely an autocratic system well managed by AI. For instance, together with strict environmental laws for individuals and businesses, the social credit system enables governments to rapidly build renewable energy capacity, reduce greenhouse gas emissions, and build cities in which all taxis and buses are electric.[71] It promises to do away with the economic inefficiency of earlier autocratic governments, such as communist systems and dictatorships. Consider Africa, where the Chinese Communist Party has supported revolutionary and anticolonialist movements since the 1960s. The West may blame China for propping up African leaders, yet Western companies and governments themselves have also paid millions in exchange for mining rights or political allegiance. Eventually, countries in Africa and also in Asia and South America will likely take loans and advice from China to install a social credit system, which will in turn permit China to set the rules and secure access to their data.[72] Some countries in Europe, such as Hungary and Poland, may also seize the opportunity.

China will also export its Great Firewall, the software that isolates a country's internet from that of other countries and determines what the people within can see and share. Tech companies will compete to get their share of contracts to build Great Firewalls around the world, as they did in the past, when Cisco helped build China's Great Firewall from American "bricks,"[73] or when Yahoo!, once the hottest tech company, launched a Chinese subsidiary that Reporters Without Borders dubbed a "Chinese police auxiliary."[74] (Similarly, when Google launched its Chinese search engine in 2006, it complied by preventing Chinese citizens from accessing forbidden information. For instance, searches for "Tiananmen Square" on Google.cn resulted in lots of tourist photos and no tanks.[75] A decade later, Facebook reportedly developed censorship tools that prevent posts from appearing in users' news feeds in specific countries to help the company regain entry to the Chinese market.[76]) In the second phase, tech companies will continue to comply with governmental surveillance. They cannot be trusted to protect users from censorship.

During this second phase, the public may learn to appreciate that the Firewall protects them from terrorists, scammers, child molesters, and other dangers from the outside world. People will still be freely able to discuss political problems, political leaders, and corruption, insofar as they know about these. The primary aim of the Firewall, combined with the social credit system, is not to keep information out but to prevent the emergence of group solidarity against the government. The spread of internet recruitment policies of the Islamic State and other terrorist organizations— alongside international solidarity movements such as Occupy or Fridays for Future—will become history.

People will enjoy new levels of convenience, including a single app, a superior form of WeChat, that allows a wholly mobile lifestyle. All payments, utility bills, business transactions, doctors' appointments, entertainment, selfies, and one's entire social life can be conducted using a single tool. This app also provides data for individual social credit scores. Those with high scores are granted easy access to loans, upgrades on flights, fast-track visa applications, and preferential treatment at hospitals. The social credit score has also become the currency of social life. People post their scores on their dating profiles and reduce time spent reading blogs on political opinions that diverge from the mainstream, playing video games, watching porn sites, and on other distractions because this behavior lowers their score. Those who don't adapt will be isolated and denied services; their friends on social media will unfriend them because contact with low scorers brings down their own score. In contrast, the highest scorers are the new celebrities, admired by all. Everyone can see their friends' social score online, and even perfect strangers' scores can be checked out by pointing one's phone at them on the street. It has become easy and convenient to know whom to trust.

Smart homes and cities provide further convenience. All monetary transactions are by mobile payment, resulting in a reduction of corruption and crime and an increase in trust. Mass surveillance will create substantial amounts of erroneous data and false accusations, but once the system is in place, people will learn to live with these problems and work around them. Newspeak will assist by changing minds. Surveillance is safety. Freedom is danger. Eventually, most citizens will think of their new autocratic world as just and convenient, as economic progress and security, not as a loss of freedom.

Phase Three: Will Democracy Survive?
In the third phase, the remaining democratic governments face competition with autocratic systems all over the world whose social credit systems closely watch, reward, and punish their citizens. Many of these democracies will find themselves surrounded by countries whose citizens trust their governments considerably more than citizens in democracies do. Countries once considered backward now implement decisions more quickly, be it for environmental protection or dealing with a pandemic, and circumvent the endless debates and compromises characteristic of a democracy. Checks and balances, such as the separation of powers and free media, are considered passé. No energy is wasted by fights and rivalries between political parties, no endless debates with committees and interest groups. No longer are election outcomes influenced by big nudging, fake news, or bots. After all, as a Chinese friend noted, a well-managed government that has its citizens willingly under control would not have to clean up the mess created

by voters who favored Donald Trump, Brexit, or other belligerent leaders and populist issues. AI provides the fuel for new autocratic systems that dispense with privacy and freedom and, at the same time, make their citizens efficient and happy. B. F. Skinner's dream of a society where the behavior of each member is strictly controlled by reward has become reality.

Eventually, the dream of the internet upholding freedom will morph into a marriage of social credit systems and surveillance capitalism, a powerful blow to the ideals of privacy, dignity, and democracy. The new ideals are fast access, high credit scores, and faith in government. In this new world of technological paternalism, laws are automatically enforced by computers. People can't cheat on taxes anymore, corruption is immediately detected, drivers won't speed and be rude to pedestrians. Surveillance brings justice, law and order. Many people will like it.

Democracies will have to make a choice. One option is to abandon the original ideal of the internet as a place for free information and adopt the social credit system and Great Firewall established in other countries, until democracy becomes a thing of the past. Another option is to stick with this ideal while facing the fact that democracies have tolerated the growth of tech monopolies that have abused and jeopardized the ideal to make huge profit. Doing so requires governments with enough courage to bare their teeth at tech companies and with a vision of how to build an internet from scratch that defends rather than dilutes democratic ideals. In other words, to go back to the original concept of the internet that is based on free exchange of knowledge rather than on surveillance capitalism.

9 The Psychology of Getting Users Hooked

We can no longer afford freedom.
—B. F. Skinner, *Beyond Freedom*

We call these people users and even if we don't say it aloud, we secretly wish every one of them would become fiendishly hooked to whatever we're making . . .
—Nir Eyal, *Hooked*

When I was a student of psychology, one of my professors held a lecture on Skinner's theory of operant conditioning. After class, the students huddled together and designed an experiment to test whether we could use the theory to control his behavior. The professor had a habit of wandering back and forth across the front of the classroom while speaking. During the next lecture, when he walked to the right side of the room, the students sitting there nodded approvingly at what he was saying. When he walked to the left side, the students sitting there showed no reaction at all. Shortly afterward, the professor spent more time on the right side of the room. Then we reversed the positive reinforcement so that the students on the left began to nod their heads and the ones on the right remained expressionless. After a while, the professor began to spend more time on the left side. Honoring experimental ethics, we disclosed the purpose of the experiment to him after class, and it turned out that he had not been aware of our manipulations. Yet we had managed to control his strolling simply through our nodding and smiles.

The logic of operant conditioning—also called instrumental conditioning—is simple and powerful:

behavior → positive reinforcement → increased frequency of behavior

It looks trivial, but its worldview is profound. Your behavior is controlled from the outside, by the positive reinforcement you receive from a human or a machine. The professor didn't stroll to the right because he wanted to; his students modified his behavior by nodding.

This clashes with our feeling of being in control and with most theories about human nature. These assume that behavior is caused by an inner state, called desire:

$$\text{desire} \rightarrow \text{behavior}$$

In psychology, desires are also called preferences and needs. In the 1940s, the American psychologist Abraham Maslow arranged needs into a pyramid-like hierarchy, with the more fundamental ones at the bottom. The pyramid includes physiological needs such as food, sleep, and sex in the bottom layer, moving up a layer to safety needs such as health and financial security and then to needs for social belonging, such as to family; at the top of the pyramid are self-esteem and self-actualization, such as becoming the ideal parent, artist, athlete, or scientist. The view that desire causes behavior also underlies Western ideas about free will. I do what I do because I want to do it, and no one can stop me unless they put a gun to my head. It is the principle underlying most legal systems: people are responsible for what they do and can be sentenced to jail, unless they were drunk or underage.

According to Skinner, all that is an illusion. Skinner, who spent much of his career at Harvard, is one of the most famous psychologists ever, and one of the most controversial to boot. Although we love to believe that we're sitting in the driver's seat, he argued that our behavior and desires are determined from the outside. Worse yet, we do not even notice that we are only in the backseat of our lives, or at least do not want to know this.

Skinner Boxes and Pigeon Warriors

Long after this experiment, when I was a visiting scholar at Harvard, my office happened to be next door to Skinner's. It was a few years before his death, a time when, thanks to an army of critics, his star had lost some of its luster. He had given up his floor of experimental rooms in William James Hall, once filled with blinking apparatuses, tons of pigeon food stacked in the corners, and pigeons confined in "Skinner boxes" with slide projectors and videos for stimulus presentation. Each box housed a lonely pigeon, whose behavior—such as the frequency of pecking at a key to get food pellets—was shaped by a fully automatized program of reinforcement.

Skinner also appeared a bit lonely, and we ended up having tea and cookies together several times. He told me that evolution and operant conditioning shape humans' and other animals' behavior and that the only difference between humans and animals is the speech muscle; our behavior is shaped by the same rules. And he proudly described how, during World War II, he had taught pigeons to guide air-launched missiles to their targets, enemy war ships. The pigeons were positioned inside the nose of a missile, where they could see through a glass, and had been conditioned to peck at pictures of ships, thereby guiding the missile toward the target. Before the pigeons could begin their career as kamikaze warriors, however, Project Pigeon was canceled. Skinner also articulated his deep belief that freedom is but an illusion, and a dangerous one at that, instilled in us by people who already control our behavior. In his view, we cannot afford the freedom to go to war, destroy the climate, and mistreat other people. Operant conditioning of proper behavior, not freedom, is the path to a better world.

In our conversations, I found Skinner to be one of those influential scholars who have developed one big system into which they squeeze every idea encountered. His techniques work well, particularly in situations where one has control over others, be it pigeons enclosed in a box or humans in similar situations. But they are not the sole key to human nature, or to making the world a better one. Today his philosophy reminds me of the promises now made by some tech companies and advocates of big nudging that, to better the world, all we need to do is give them access to all our data, turn our home into a smart box, and let them surveil and modify our behavior. That is not just an analogy; the techniques are astonishingly similar.

Intermittent Reinforcement

Back to my professor. When the students stopped nodding, the professor's behavior gradually faded away. How could they have made the professor continue to spend more time on the right side of the classroom? How quickly the behavior faded was under the students' control: if they had nodded every single time the professor was on the right side of the classroom and then suddenly stopped, he would also have soon stopped spending most time on the right side. This loss of external control is called *extinction*. However, if students had only occasionally nodded when the professor moved to the right, then his behavior would have continued much longer. This technique is called *intermittent reinforcement*. It means that reinforcement of the desired behavior is irregular, not constant: not every time, or every second or third time, and so on. Reinforcement is not predictable. Sometimes the behavior is reinforced, sometimes it isn't.

behavior → intermittent reinforcement → increased persistence of behavior

Intermittent reinforcement is the way to build up lasting behavior.[1] The hungry pigeon in the Skinner box pecks more frequently and faster when it gets a food pellet not every time but at irregular intervals. Similarly, irregular reinforcement can be used to develop persevering and hard-working behavior but also to get people hooked on smartphones or slot machines.

How to Get Users Hooked

Assume you are a software engineer, and your company pays you for increasing customer satisfaction. Its business plan is "users pay with their data," and your customers, say the advertisers, want users to spend as much time as possible on an app for maximum exposure to their ads. How would you proceed? The answer is to program intermittent reinforcement. This is how it works:

Step 1: Identify the behavior you want to increase: time spent on a site, number of clicks, click-through rate, number of posts, or something else.

Step 2: Identify a positive reinforcement that can build up and control this behavior. Social media networks have no food pellets, drinks, or hugs to offer as reinforcement. But they can use another currency: social approval and the dopamine shots one gets from being accepted as a member of a group. Thus, social approval is a potential reward to make users spend more time on a site.

Step 3: Divide the reinforcement into small, countable units. Social approval works best for getting users hooked if it comes in small, distinct units, like the food pellets in Skinner's experiments.

Step 4: Introduce intermittent reinforcement. This spreads reinforcement in an unpredictable way over time and keeps users watching, clicking, or posting.

So what positive reinforcement could you introduce? Nodding would be difficult. But think about nodding as notification or, even better, likes. These come in distinct units and at irregular times.

The Like Button

In the first versions of Facebook, little existed to bind users to the site. People sent their messages and that was it. In later versions, comments on others' posts were possible, which acted as a reward for checking one's site. Positive comments are reinforcers. The real breakthrough came when the Like button was introduced in 2009. Unlike comments, which may have sarcastic undertones and require evaluation, the Like button resolved all ambiguity. It was even simpler than a number between one and ten. A like is a like, just as a food pellet is a food pellet. It is even possible to count them, along with the number of followers, which makes social comparison as simple as comparing two numbers. Likes do not necessarily follow every act of checking in; reinforcement is intermittent. The like became the ideal unit for reinforcement that, together with comments, could control users' behavior:

checking social media → intermittent reinforcement by likes → increased frequency and persistence of checking

In this way, several behaviors—checking, scrolling, clicking, and post-ing—are influenced by likes. The impulse is to check constantly because one never knows when the next like will come. This behavior is almost automatic; people continue to scroll and click, just once more, even when they believe it's a waste of time. Day after day, millions of people spend hours posting pictures on Instagram, waiting expectantly for feedback, and counting the likes they get. Likes become hugs. Several years ago, in fact, designers at MIT Media Lab made headlines after developing a smart vest that gives you a hug for each like you get on Facebook.[2]

One might assume that if a user gets a large number of likes, they might slow down and enjoy their fame. Skinner's theory instead predicts that the more likes a user gets, the more frequently they will post, which decreases the time between successive posts. That is exactly the pattern that was found in an analysis of over one million posts on Instagram and other social media platforms.[3] The number of likes controls users' frequency and timing of posting. Other studies concluded that the brain regions activated by likes closely overlap with those activated by nonsocial awards such as food, just as with Skinner's food pellets.[4]

Before email and social media, one of the ways people maintained contact with others was by writing letters. The letter carrier delivered the letters at a certain time of day, say at noon, so people checked their mailbox once a day. This corresponds to what is called a *fixed-interval reinforcement* schedule, which leads to one spike of behavior and then nothing until the next day. Many a person expectantly peered out the window with anticipatory joy, waiting for the delivery of letters or packages, but once they arrived, that was it for the next twenty-four hours. Messages via email or social media sites, in contrast, can arrive at any time, 24-7, and arrive irregularly. This design produces intermittent reinforcement, which leads to constant checking behavior. It would correspond to a postal system in which several letter carriers might turn up at any time of day or night, carrying only one letter or package each.

It might seem that Facebook deliberately went through steps 1 to 4 above and designed the Like button in response. Yet that was not so. As with many discoveries, Facebook appears to have serendipitously stumbled over the Like button while trying to solve another problem. When the site grew in popularity, posts were overrun with comments stating little more than that someone liked it. The Like button was originally intended as a means of cleansing the site from this wave of redundant comments. Only later did it become clear what a treasure they had unearthed.

The Tools of Attention Control Technology

Likes are the glue that keeps people hooked. But they don't act alone. Social media sites run experiment after experiment to find out how to keep users bonded with the screen. Sean Parker, the first president of Facebook, explained the goal of social network platforms: "How do we consume as much of your time and conscious attention as possible?" To do so, he continued, whenever someone likes a post, we "give you a little dopamine hit" to encourage you to upload more content: "It's a social-validation feedback loop . . . exactly the kind of thing that a hacker like myself would come up with, because you're exploiting a vulnerability in human psychology."[5] Dopamine is one of about twenty major neurotransmitters that carry urgent messages between neurons, nerves, and other cells in the body— like bike couriers weaving through heavy traffic. It becomes active when anticipating or experiencing rewarding events, such as when we bite into delicious food, have sex, and receive social approval.

Here are a few ingenious techniques to capture users' attention and time, some of which are so common that we don't notice them anymore.

News feed Sites like Facebook began as a directory of profile pages on which users could list their favorite bands or post pictures. Users visited their friends' home pages from time to time to view their updates, which were relatively minor. In 2006, when Facebook was two years old, it announced a facelift called News Feed. The announcement bore the smell of its dorm room origins: "It updates a personalized list of news stories throughout the day, so you'll know when Mark adds Britney Spears to his Favorites or when your crush is single again."[6] Hundreds of thousands of users protested against this new technology that turned their private messages into the fodder of mass consumption.[7] But technology won and replaced users' value of privacy with the desire to receive as much attention as possible, not only from friends but from anybody on the globe. What the Facebook announcement neglected to mention, however, was that News Feed did not show all posts but used an algorithm to select them. The first algorithms were based on intuition about what people liked, giving five points for a photo or one point for joining a group, and multiplied this score by the number of friends involved in the story.[8] With the help of the Like button, which provided a direct measure of what people liked, this crude algorithm eventually morphed into a top-secret machine-learning system that determines what content a user can easily see (because it is placed near the top of the News Feed), what content is hard to find (placed far away from the top), and what content the user will not see at all (because the algorithm decided the content is not relevant for the user). As a result of the News Feed, users spend more time on the site.

Notification systems Likes feed into a notification system, which is a control feature to bring users back to the site. The moment someone likes a post, a comment, or comment within a comment, an alert appears that can be clicked on. Over time, Facebook increased notifications; the more notifications, the more frequently people check in.

Delay likes Notification algorithms have been reported to sometimes withhold likes so that the person who makes a post initially does not get any likes, or only a few, and may be disappointed. Only later are the likes sent in larger bursts. This delay amounts to an amplification of intermittent reinforcement to make people stay on even longer.[9]

Autoplay   In a movie theater, you pay for the movie, watch it, and leave after it ends. On YouTube, after you watch a video, a related video automatically begins in order to keep you watching and exposed to as many advertisements as possible. According to YouTube, the average session on a mobile device lasts more than an hour, and 70 percent of all watching time is controlled by the recommender system, including autoplay.[10] People make few choices themselves. Not only does YouTube's recommender algorithm lure viewers into spending more time on the screen longer than intended, but it can also lead them toward extreme and unscientific viewpoints.[11] That is, the fewer choices people make, the more likely they end up with untrustworthy information.

Snapstreaks   A streak is the number of consecutive days a user and a friend have sent each other a "snap," that is, a photo or video. Platforms such as Snapchat display the number together with icons such as a fire emoji. The hook is that the number drops to zero if a day is missed. This feature nudges users to keep responding to their friends; not doing so can cause disappointment and anger. As a consequence, users tend to send reminders to their friends that they are awaiting a reply. Diehard users have continued their streaks for more than 2,000 days.[12] Snapchat also awards points for every snap sent or received. A low score is considered embarrassing. And there are virtual trophies, such as specific emojis for sending specific kinds of snaps. All these features keep users sending snaps and spending more time on the platform. The average Snapchat user sends thirty to forty messages every day.[13]

Mindless games that require constant attention   In the Facebook game *Farm-Ville*, players have to tend a farm. There are set time limits, such as if you don't return to the game within sixteen hours to harvest the rhubarb, your fields will be riddled with withered stalks.[14] In 2010, the game boasted more than 80 million monthly active users, over 30 million of whom were playing daily. Some spent not just their time but also their money on virtual farming (the game is free, but if you purchase certain features, you are able to farm better). *Time* named the game one of the "50 Worst Inventions" in recent decades for consisting of mindless chores on a digital farm and for having the most addictive design among Facebook games.[15] The contents of an addictive game are irrelevant. Video game designer Ian Bogost learned this lesson when he created a parody of *FarmVille* to lay bare its mindless and repetitive character. The game was called *Cow Clicker*. Players could click on a picture of a cow, the cow said moo, and players earned another click in order to be able to click on a cow again in six hours. Clicking earns clicks; that was it. The six-hour span kept the players returning to the game, although the time could be reduced for a fee. Players could invite friends to click on their cow, and when they did, they all received another click. Each time they clicked on a cow, this was announced to their friends on Facebook. The message read: "I'm clicking a cow." Bogost had expected *Cow Clicker* to have a short life because of its ludicrousness. Instead, within weeks it achieved a cult status, and the number of players grew to tens of thousands. In the end, Bogost shut the game down in desperation, an act he called Cowpocalypse.[16]

Loot boxes Many video games offer users the option to buy virtual loot boxes, which may contain useful resources, such as weapons and armor. Like the buyers of real loot bags of old, users find crap in the box most of the time, but occasionally they find something really good. This intermittent reinforcement fosters excitement and surprise and keep users purchasing more boxes. A survey of more than 7,000 gamers found evidence for a link between the amount spent on loot boxes and the severity of their gambling problem.[17] The UK's National Health Service warned that this randomized reward structure sets up kids for gambling addiction, and some countries have regulated or banned the sale of loot boxes.[18]

Each of these techniques is designed to hook users to a site: to make it difficult to leave a platform and compelling to return to it. For many of us, the effect is a mixture of feeling good, a favorite pastime, and a waste of time and attention. For others, the result is addiction: a feeling of loss of control and a constant craving for positive reinforcement.

The designers of these attention techniques are well aware of the consequences, as are the heads of digital companies, who place strong restrictions on their own children's internet use. As Tim Cook, CEO of Apple, explained: "I don't have a kid, but I have a nephew that I put some boundaries on. There are some things that I won't allow; I don't want them on a social network."[19]

The Power of Your Smartphone

You might object that paying attention to your smartphone is always a deliberate decision, for instance, because you're expecting a message. And if you don't want to bother with messages, you simply turn the phone off. You feel in control. But are you?

This situation was tested in an experiment with over 500 undergraduates.[20] One group was asked to leave all of their belongings in the lobby, including their smartphones, before entering the test room. A second group was instructed, after entering the test room, to put their phones away in their pockets or bags. A third group was instructed, after entering the test room, to place their phones face down on the desk, for use in a later study. All participants were asked to turn off the ring and vibration settings. Then they were all given standard tests of attention and intelligence, such as completing a series of math problems while simultaneously remembering a letter sequence. Because nobody's smartphone could ring or vibrate, its location should have had no impact on the students' performance. Yet one group performed better than the others in every single task. It was the group whose smartphones were in the other room. Those who had their smartphone on the desk performed worst. The nearer the smartphone, the stronger its claim over the owner.

Would that still happen with smartphones that are not only silenced but turned off completely? In a new version of the same experiment, turning the smartphone off made it no less distracting. Its mere presence was again enough. After the experiments, the students were asked whether they thought the location of their smartphone had any impact on their performance. The far majority believed that it did not. They were unaware that the phone controlled their attention. Ironically, those who reported they would have trouble getting through a normal day without their phone suffered most from its brain drain.

This study suggests that the mere presence of students' mobile devices in schools and universities, *even when they are not activated,* may undermine attention, learning, and test performance. Unless the devices are needed in class, leaving them in a different room is the wisest strategy for learners.

Smartphones are not the first technology to demand our attention; TV, radio, and the regular phone have been accused of doing the same. My point here is that smartphones can reduce the ability to concentrate even when they are silenced or shut off, and that few are aware of this phenomenon. The problem is not technology; it's whether we understand its impact on us and whether we learn to regain the upper hand. Many are more intimately attached to their phone than to their friends and family. In a survey of 500 adults in the United States, two-thirds said they sleep with their phone at night, and almost half said they would rather give up sex than their cell phone for a year. The far majority—three-quarters—consider themselves addicted to their phone.[21] The result is a permanent state of distraction, of not being able to pay full attention to other tasks and use one's intelligence at full capacity. If you need to concentrate on something else, put the phone away—far away. As an old saying goes: out of sight, out of mind.

Is Social Media Addiction Like Gambling Addiction?

Throughout her pregnancy, Isabella, a thirty-eight-year-old living in Las Vegas, had gambled, losing every dollar she owned. She continued until days before giving birth to her son. For her evening therapy group, she wrote about how most days she would spend some sixteen hours in a chair in front of a slot machine, before and after giving birth, with people smoking around her. She barely ate a thing because she did not want to lose any time at her machine:

> Even after he was born I couldn't stop gambling. I'd leave him at home with my sister for hours and hours. Later, after losing everything, there would be stains all the way down to my hips from the leaking of my breasts. He'd be at home hungry, and I'd be gambling it all away. Now I'm trying to stop gambling, but I get so bothered by the machines when I get baby formula at the store. I try to close my eyes and get past them, but it doesn't always work.[22]

In Las Vegas, slot machines are not only waiting in casinos but also beckon players at the entrances to gas stations, supermarkets, and drug stores. Isabella could hardly avoid the sight of them.

Is addiction to gambling similar to addiction to social media? There are indeed similarities. The first is the goal of the companies that provide the machines. In her brilliant book *Addiction by Design*, Natasha Dow Schüll describes the efforts of slot machine engineers and developers to keep people glued to the machine. As game designer Nicholas Koenig explains, the hidden algorithms are critical for getting players addicted: "Once you've hooked 'em in, you want to keep pulling money out of them until you have it all; the barb is in and you're yanking the hook."[23]

To meet this goal, slot machines have become "personalized." Two machines that look similar from the outside may have different algorithms that fit different customer types. For instance, *escape players*, who want to escape from the uncertainties of their lives by spending as much time as possible at the machine, will likely get hooked on machines that are programmed to dispense little payouts at random points in time. *Jackpot players*, in contrast, are not interested in little payouts. They are willing to lose big in order to win big and prefer machines with rare but high pay-offs. For escape players, video poker has become the favorite game, where they can spend about two hours on the machine for every $100 lost, about double the time that can be spent on regular slot machines for the same price. That, by the way, is more expensive than recreational drugs.[24] Reinforcement occurs more frequently but, as with the food pellets in Skinner's experiments, in small units, which maintains a high frequency of gambling behavior. The gambling industry's term for this business model is time on device (TOD).[25] Thus, the common business model of both slot machine platforms and social media platforms is to capture as much of a person's time as possible.

The second similarity is the use of intermittent reinforcement. It enables social media to keep people scrolling though tweets and clicking through photos of friends of friends, even if there is nothing particularly interesting and they know they will have a "lost time" feeling afterward. People also grow addicted to slot machines through unpredictable shots of dopamine. Wins come after a random number of games, which leads to a steady and high frequency of playing. As Isabella perceived, the machines exert control over her behavior—they make her sit down and play. Intermittent reinforcement is powerful, but some people more easily fall prey to it, depending on other influences. In the case of Isabella, there was a history of an abusive, alcohol-dependent father and additional factors over which she had no control.

But there is also a difference between gambling and social media. Slot machines resemble the boxes in which Skinner trained pigeons, a lonely bird allotted to each box. Intermittent reinforcement keeps the pigeon pecking at a key and the gambler pushing the buttons, which sometimes leads to a pellet or money as a reward. People addicted to social media find themselves in a different situation: the very point of social media is being connected, gaining social approval, and making social comparisons. The likes provide an easy currency to see the extent to which one is accepted; a post with zero likes is equivalent to public condemnation. You might have asked yourself why the people who broke into the US Capitol in 2021 posted selfies and thereby publicized identification photos that led to their arrests. They may not have been simply dim-witted; many were perhaps continuing their perpetual quest for the positive reinforcement of likes. To post means to be seen; to play a slot machine does not.

Addicted gamblers are often ashamed of their addiction. They play alone and do not communicate with others. Even when gamblers sit next to each other, there is little if any social interaction. In one case, a surveillance camera was accidentally directed at a man at the tables who suddenly collapsed onto his neighbor, in the throes of a heart attack. The unconscious victim was literally lying at other players' feet, his body touching the bottoms of their chairs. Yet all of the neighboring gamblers continued to play, without batting an eyelid.[26]

Many among us find it hard to understand why Isabella and others like her spend their lives in the machine zone. Some researchers think these people cannot calculate probabilities. That explanation misses the point: experienced gamblers are highly sensitive to probabilities, even to differences between personalized machines that look the same. Others believe that gamblers are driven by the urge to win. Yet addicted gamblers do not necessarily play to win. They play simply in order to spend time with the machine. Their behavior is under the sway of the machine's intermittent reinforcement algorithm. As one woman explained, when she wins on video poker, she just puts the money back into the machine. "The thing people never understand is that I'm not playing to win." Why does she play? "To keep playing—to stay in that machine zone where nothing else matters."[27]

10 Safety and Self-Control

Not being able to control events, I control myself.
—Michel de Montaigne, *Essays*

Control yourself or someone else will control you.
—Anonymous

A mother addicted to online video poker is under the control of the software's intermittent reinforcement, which makes it hard for her to leave the site to spend time with her family and may empty her bank account. A young man eager for attention turns to increasingly violent behavior, fueled by positive reinforcement on various social media platforms.[1] Staying in control, or getting it back, is not always easy. Ever since Homer recounted the story of Ulysses, who asked his sailors to bind him to the mast to avoid being lured by the sirens' sweet but deadly song, people have thought of numerous ways to stay in control. Some make a pledge to their beloved late mother or father, or to a friend. Others turn to Gamblers Anonymous or phone addicts anonymous, or find a buddy with whom they can together limit the time spent on online gambling, internet porn, or shooting games designed to be addictive.

Staying in control is not only difficult, it is made more difficult by technological design. Similarly, cigarettes have been designed to be addictive by the addition of certain chemicals that are redundant for the mere pleasure of smoking, and slot machines are designed to keep gamblers playing even if they would feel better if they could stop. The tobacco industry, the machine gambling industry, and tech companies have all been asked by health organizations to stop exploiting human vulnerability. And, to some

degree, tech companies have complied by promoting apps that allow notifications to be turned off, limit who can see the likes, or remind users to
take a break.

In this chapter, we look at how to foster digital self-control. Self-control
does not mean staying clear of games and snubbing distractions. It means
being able to stop when one would actually prefer to be doing something
else, knowing that one will regret the time spent that way afterward, or
when the activity threatens one's health and that of others.

Distracted Drivers

Distracted driving occurs when you take your eyes from the road, your
hands from the wheel, or your mind off the task of driving safely. That happens when you reach for your phone, text while driving, check emails, take
a photo and post it, or watch a video but also when you turn to reach for
a bag of potato chips in the back seat. Multitasking behind the wheel has
become deadly normal. It kills far more people than terrorists do. According
to the Centers for Disease Control and Prevention, every single day in the
United States about eight people are killed in crashes involving distracted
drivers.[2]

The Last Text

It was Christmas Eve, and the young woman was eager to get home. But she never
arrived. With only a few miles to go, she speeded without braking straight into
an 18-wheeler. The street was dry, the visibility was good, there was little traffic,
and the firemen at the scene of the accident wondered how it could have happened. When her body was carried out of the wreck, they found an intact cell
phone lying on the floor of the car. It displayed the woman's last message: "I'll
be right back."[3]

It's 8:14 p.m. on June 21, 2014, and Laura is driving her Mazda 3 north on Route
de la Station in L'Isle-Verte. She sends a text message to her friend. Two minutes
later, a reply comes buzzing on Laura's iPhone and she opens the messages while
the car approaches a rail crossing at the crest of a hill. The crossing's red lights
flash, its bells clang and the oncoming locomotive sounds its whistle four times.
But Laura never slows down. The train barrels into Laura's car at 64 kilometers
an hour.[4]

Hi. My name is Jenna. Some could say I don't really have a story. I've never lost
anyone close to me to a texting driver. In a way I do have a story. I was in the car

a few weeks ago, when I saw a man in a red car texting while he was driving beside us. He suddenly swerved and nearly hit us. He almost hit my side. My father had honked furiously, he almost lost his only daughter. And you know the worst part? After the man looked up with an apologetic look on his face, he went right back to texting! I bet he killed someone that day. Stop texting and driving, I'm only 12. I'm too young to die.[5]

Typically, the victims of distracted driving die for no good reason. The distracting messages sent and received are trivial. People die because of the urge to reach to the phone, which is controlled by intermittent reinforcement. How can we regain control? One way is to take half an hour and listen carefully to those who were unlucky. Years ago, AT&T approached the legendary German filmmaker Werner Herzog to make a documentary about texting and driving. Herzog agreed and created the haunting movie *From One Second to the Next.* Victims and wrongdoers openly talk about the disaster they experienced or caused. A mother whose son has been paralyzed recounts how their lives have been destroyed, as does a young man who killed three little children while texting "I love you" to his wife. One case Herzog did not get the rights to cover was that of a young man who was writing a message to his girlfriend and steamrolled a child on a bike. His girlfriend was sitting next to him in the passenger seat.[6]

Apart from being shown in numerous high schools, *From One Second to the Next* can be seen on YouTube. After watching it, I looked at the viewer count. It was a low five-digit figure. By comparison, "How to tie the perfect bow tie" or "How to apply eyeliner" videos easily get millions of views.

Multitasking

Multitasking means performing two or more tasks at the same time, each of which demands attention. Texting while driving is an example. Breathing while driving is not because breathing happens automatically. Attention is the crucial limited resource. The *fundamental law of attention* is this:

> If you perform a task that requires attention and then simultaneously perform a second one, your performance on the first task deteriorates.[7]

Multitasking is no big deal when mowing the front lawn while listening to a podcast or watching a movie while chatting with a friend. But in a car, we are operating a potentially deadly weapon. Despite the steady stream of victims, many drivers are of the opinion that the fundamental law of attention does not apply to them because multitasking has become second nature. Psychological studies show that this popular belief is unfortunately an illusion.

For instance, one study compared frequent multitaskers with rare multitaskers. Those who were used to multitasking were found to be more easily distracted by irrelevant information, had a worse memory, and were slower at changing between tasks.[8] All of these are abilities at which seasoned multitaskers should excel. But aren't there at least a few exceptional people who are not subject to the law of attention? That question was addressed by a study in which students had to solve simple memory and math problems using a hands-free headset in a driving simulator. As usual, their performance in four tasks—brake reaction time, keeping distance, memory, and math—dropped in comparison to their performance in each single task. An individual analysis showed that this result held for all 200 students, except for five who were reported to have "absolutely no performance decrements in multitasking" and were celebrated as "supertaskers."[9] Are they really? When checking the study, I found that, contrary to what was suggested, four of these five supertaskers actually performed worse in one of the four tasks. That leaves one out of 200 students. Yet that individual might well be a chance oddity. For instance, if you ask 200 people to roll a dice four times, it is not unlikely that one will have happened to roll six four times in a row. That alone does not prove that someone is immune to the laws of chance; this "supertasker" would need to be further tested to rule out chance.

All in all, a large amount of research has shown that attention is a limited resource and that performing additional tasks that require attention lowers the quality of performance on the original task. Practicing multitasking cannot overrule this law of attention and turn us into supertaskers.

Hands on the Wheel, Eyes on the Road

Why do people text while driving? If you ask teens, some may retort that adults do it too. And they are spot on. In a US national survey, parents were asked about texting and driving. Half were mothers, the other half fathers, all of whom had driven their own child in the previous thirty days.[10] When asked, "Do you think that you can safely text and drive?," most said "never." But when asked about their actual behavior, most admitted that they had read and written texts while driving in the past month.

Intermittent reinforcement and shots of dopamine have taken control over their behavior against better understanding. To assist parents, pediatricians and nurses should routinely ask them about texting while driving their child. Yet only one out of four parents in the survey reported that their pediatrician had ever broached the topic.

As mentioned, distracted drivers kill about eight people in the United States every day, orders of magnitude more than terrorists do. In spite of this daily death count, many drivers continue to text on the road. Is there a cure for this? One way is to resort to strategies such as putting the phone out of sight and reach when in the car. There are also websites where individuals can officially take the pledge that they will pay full attention to the road and not use a phone while driving or give in to other distractions. Many people have resorted to digital technology itself to counter their urge to text on the road and purchased an app that detects driving, blocks all incoming and outgoing messages, and responds with automated messages without alerting the driver. In the national survey, one in five parents reported using such a self-control app. These apps are safer than smartphones that text via voice command. The latter allow you to keep your hands on the wheel but take your mind off the road. They can provide a false sense of safety.

Safety issues are not limited to the road. While preparing for landing in Singapore, the crew of a Jetstar Airbus A321 heard noises associated with incoming text messages from the captain's mobile phone.[11] The first officer, who was flying the plane, repeatedly asked the captain to complete the landing checklist, but the captain was preoccupied with his phone and did not respond. The aircraft had to make a go-around because the landing gear was not deployed in sufficient time. Distraction due to texting has also been reported as contributing factors to helicopter crashes in Florida and Missouri. In Colorado, the pilot and the passenger of a Cessna 150 took selfies using a flash at night, and the National Transportation Safety Board concluded that this action contributed to the subsequent crash that killed both occupants. In response, several countries have prohibited pilots from using electronic devices for personal reasons while on duty.

Distracted Parents

One day, a Yale graduate student in economics took his son to the playground. The very moment he looked down to check his phone, his son fell. It wasn't a serious accident, but it provided the young economist with a hypothesis. When AT&T rolled out its 3G networks across the country in 2008, he was given an opportunity to test it. AT&T rolled out its service at different times in different regions, which allowed for a natural experiment. Region by region, as smartphone adoption rose, he found that emergency units reported an increasing number of injuries for children under five, such as broken bones and concussions.[12] That was consistent with what happened at the playground but could of course be caused by something other than distracted parents. The economist set out to find out who or what was driving these injuries. He found that the increase happened when parents watched their children at public playgrounds, at pools, or at home, but not when teachers or coaches supervised the children. Similarly, the German Lifeguard Association issued a warning that an increasing number of parents are fixated on their smartphones instead of supervising their children on pools and beaches, risking the hazard of a child drowning before they notice.[13] Drowning can be fast and silent.

The sight of parents distracted by their mobile phones in the company of their children has become normal. Like the Yale student, parents sit on playground benches early in the morning, scrolling and clicking, absent to the fact that their little child has no one to play with. Others push buggies while gazing at the phone screens rather than at their babies seeking to make eye contact with them. Children complain, beg, and try to make the phone go away. As the story goes, a five-year-old hid his mother's Black-Berry so that she would talk to him, and another boy got so upset that he flushed his father's iPhone down the toilet. But not much later, the same parents will likely complain that their children themselves are glued to the black mirror.

Sean Parker, the first president of Facebook, once commented on the use of Facebook: "God only knows what it's doing to our children's brains."[14] Parker was talking about children's excessive use of social media. But parents' use can have a similar impact on their children's development. Leaving aside the question of how harmful excessive use of digital media actually is, there is an undeniably direct effect: the more hours parents spend on the phone, the fewer hours are available for parenting and bonding. What this loss of time together does to children's development is less well understood.

The result can be physical damage, such as noticing too late that children are about to injure themselves, and psychological damage, such as children developing feelings of not deserving attention.

Half of US teenagers said that their moms or dads are distracted by their phones while the teenagers try to talk to them, and many wish their parents would spend less time on the phone. One in four teenagers believe that their parents are addicted to their phones.[15] Teens with distracted parents also feel that their parents express less warmth, a possible cause of teen anxiety and depression. As mentioned before, however, these studies report correlations, not causes. Parents appear to notice when using a phone that their children are not as relaxed and more easily upset, but some continue anyhow. These children get the signal that whatever their parents do on their smartphone is more important than they are.

Conversely, when parents watch what children do, this sends a signal to children that what they are doing is important, and they try harder. Children even perform better at sports when their parents are watching than when parents are on their phones. In general, children need attention in order to develop their skills. That is also the key to early language learning. In one study, thirty-eight mothers taught their two-year-olds two novel words, with one minute of teaching time for each word. One of these periods was interrupted by a cell phone call, the other not. If the period was interrupted, the remaining time was always added afterward, so that the total time spent on teaching each word remained identical. When the teaching was not interrupted, children learned the word; when teaching was interrupted, they did not.[16]

Studies also show that the best parents can do for early language development is to read stories aloud to their children and keep going even after a child can read. In contrast, when watching baby videos such as *Baby Einstein* and *Brainy Baby*, children under the age of two learn less than by listening to a person.[17] These allegedly smart early language-learning programs are designed primarily to grab kids' attention, and they do not deliver what they promise.

Daddy, Where Does the Phone Sleep?
Intermittent reinforcement makes quite a few people check their phone more often than they say they really desire. For some, this urge cannot be

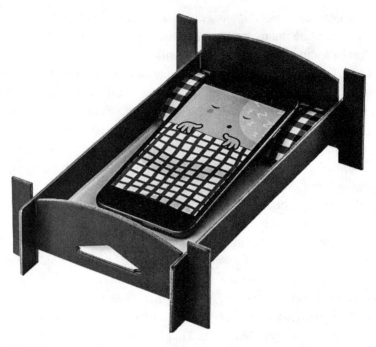

Figure 10.1
Smartphone bed. Children learn the habit of putting their phone in its little bed before getting into their own. While "sleeping," the smartphone—like a human being—is loaded with energy. Shown is the original German version, distributed by the Auerbach Foundation.

resisted during yoga, a religious service, or a funeral. A dopamine shot may become more important than the goal to relax, to pray, or to commemorate a deceased friend. To regain self-control, building up new habits may help. Research indicates that self-control measures attenuate the relationship between use of electronic media and loss of sleep.[18] And these should be learned as early as possible. Some parents have agreed on smartphone-free time or space with their children; others agree on having no phone at the table during meals. Others again have bought or built a cute little bed along with their child's first smartphone (figure 10.1). Before getting ready for bed, the child can put the phone into its own little bed outside the child's room. This creates a habit that can help prevent children from spending their nights texting, posting, and watching videos under the blankets. Both phones and humans reload while asleep.

These habits and devices define the culture in a family and can be adopted to the specific needs of its members. Whatever the rules agreed upon, they should hold for parents and children equally. There are no limits to creativity. In the hallway of a friend's home I discovered a bag hanging on the wall with eight pockets, inviting guests to leave their phones there in order to foster a good dinner conversation. If you truly want to spend *less* time on the phone and *more* time on personal relationships with children, friends, and family, there are plenty of means to make this possible.

Use It or Lose It

In the early hours of June 1, 2009, Air France Flight 447 was on its way from Rio de Janeiro to Paris. It was business as usual: passengers were sleeping, dozing, reading, or watching a video. The route crossed the intertropical convergence zone, passing through the top of a powerful thunderstorm. At 2:02 a.m., after a short briefing, the captain took a break to get some sleep. His two copilots took over the plane. Twelve minutes later, the plane plunged into the Atlantic Ocean, killing all 228 people aboard.

What had happened? Four minutes before the crash, the autopilot disengaged. It disconnected because the air-speed sensors iced over, and it no longer had reliable data. The two copilots were forced to fly the plane manually, without reliable indicators of airspeed. Within a fraction of a second, the crew found itself in one of the most dangerous crisis scenarios. Neither of them had ever received training on how to fly the airplane in this situation at cruise altitude. What made it worse, at the moment when the autopilot disengaged and the plane rolled to the right, the pilot flying reacted with a correction to the left that was too extreme for high altitude and destabilized the plane. At high altitudes, the air is thin, and smaller manual corrections are needed than at low altitudes. The pilot flying also tried to put the airplane into a climb to escape from the thunderstorm, which caused it to lose speed. That initiated two loud *stall* warnings in the cockpit. Stall means that the minimal speed necessary to keep the plane flying is no longer obtained, and the plane plunges. Pilots are trained to put the nose of the plane down in such a situation to gain speed again. But the black box record showed that the pilots were confused about what was going on. The aircraft fell like a stone from the sky. One can only imagine what the terrified passengers and crew experienced in the final moments. The last recorded speed before impact was just over 120 miles per hour.

The final report of the investigation pointed out a lack of training in manual flying at high altitudes as a contributing factor to the disaster. In the course of increasing automation of planes, basic flying skills were insufficiently taught.[19] The disaster of Flight 447 had many causes, some of which remain uncertain, but one thing is clear: the crash was entirely unnecessary.

Automation for Regular Events, Humans for Unexpected Events

Airlines and plane manufacturers were among the first to move toward autopilots. Automation can increase safety in situations that are stable and predictable, as in regular high-altitude cruising, but routinely relying on it can result in pilots who have little experience what to do if the autopilot fails. The Air France Flight 447 crash and similar accidents indicate that pilots have become too dependent on computerized systems and may not know what to do when unexpected things happen. In response, the Federal Aviation Administration (FAA) released a safety alert for operators in 2013. It recommended that airlines instruct their pilots to spend less time flying on autopilot and more time flying by hand and sight.[20]

Like the FAA, the US Navy has realized that growing dependent on automation can be fatal. Around 2000, the navy had begun to phase out training its service members to navigate by the stars or use sextants and charts in favor of electronic navigation systems, such as GPS.[21] After the initial enthusiasm that perfect computer technology can replace imperfect human judgment, the navy realized that in a war, satellite signals are likely to be hacked or jammed, and that satellites may even be shot down. Meanwhile, the Naval Academy has now gone back to basics and trains its members to use their own brains.

As we saw in chapter 4, automated cars share the same problem. Computers can take over an ever-increasing number of routine tasks, such as parking and overtaking on highways, but when something unexpected happens, a human driver is needed to step in. Unlike the pilots of Air France 477, human drivers have even less time to react, often only a few seconds or fractions of seconds. This is why alert drivers are needed in regular traffic. Outsourcing driving skills to board computers and sensors does not relinquish this need. As mentioned earlier, one way out is to build closed and controlled highways or cities that are adapted to the limited capability of automation. This option, however, does not appear as easy for aviation. The general dilemma follows from the stable world principle: outsourcing navigation skills works as long as everything goes according to plan, but alert and trained humans are required for situations where something happens out of the blue. This dilemma is known as the *automation paradox*:[22]

> The more advanced an automated system, the more crucial an experienced and attentive human controller.

A similar dilemma exists with GPS systems. They are immensely useful for driving or walking. But relying on them routinely in everyday life reduces the development of spatial reasoning, including navigation ability and the ability to form a mental map of the environment.[23] When we use GPS all the time, our brain doesn't bother to build a cognitive map of our surroundings. We hardly know where the river or lake is, or where north or south is. When the phone battery unexpectedly dies, we will not know where the restaurant is where we are supposed to meet our friends, and since we have not memorized the phone numbers of our friends either, we are lost. We might not even know how to get back to our Airbnb. In a world of certainty, where GPS always works without a hitch, our spatial sense and memory could be outsourced and dispensed with. In the real world, GPS can fail. It already failed for more than twelve hours in 2015 due to failures in the satellite network.[24] It may get a bug, run out of batteries, or become the victim of an international hacker attack.

Using GPS started as an option. It helped to navigate through new and unknown territory. Overusing this option leads to a vicious circle. Previous skills—reading a map, knowing where north is, memorizing street signs, noticing landmarks—become underused or are lost entirely. That, in turn, makes people more dependent on the system. When something unexpected happens, then there is no option left. To break this cycle, it can be helpful to develop habits that allow the advantages of GPS technology to be used without losing all of one's own navigation skills. Some people use GPS only for finding new locations, but then exercise their spatial memory to find old ones. Others use GPS while walking but turn the audio off.

Trying to find old locations with your own spatial sense keeps it engaged and alive. Heedlessly following audio commands such as "turn right now" and doing precisely what one is told is the worst for fostering a spatial sense. Astounding stories have been reported about people making a trip to pick up some groceries and ending up hundreds of miles from their destination.[25] As a famous scientific study with London taxi drivers has shown, developing a sense of navigation changes your brain.[26] These taxi drivers, who had no GPS, developed alterations in their brain functions as they learned to navigate. The brain is like a muscle: it needs to be exercised. Use it or lose it.

11   Fact or Fake?

Children must be taught how to think, not what to think.

—Margaret Mead, *Coming of Age in Samoa*

Instead of seeking to outperform the human brain, I should have sought to understand the human heart.

—Kai-Fu Lee, former head of Google China

We think it is easy to tell a fact from a fake. Sometimes it is, but at other times it can be extremely tough—particularly if a narrative contradicts one's own experience. The seventeenth-century English philosopher and physician John Locke told this story:

> And as it happened to a *Dutch* Ambassador, who entertaining the King of *Siam* with the particularities of *Holland*, which he was inquisitive after, amongst other things told him, that the Water in his Country, would sometimes, in cold weather, be so hard, that Men walked upon it, and that it would bear an Elephant, if he were there. To which the King replied, *Hitherto I have believed the strange Things you have told me, because I look upon you as a sober fair man, but now I am sure you lie.*[1]

Locke was a founder of empiricism, the claim that all our knowledge, except possibly logic and mathematics, stems from experience. He distinguished between certainty (knowledge) and probability (reasonableness). If you have seen a man walking on ice, this is knowledge, not probability. If you haven't, it is probability. For Locke, probability depends on two factors: conformity with one's own experience and the testimony of others' experience. If the King of Siam, who has never seen water turning into ice, is told that the Dutch can walk on it, this report contradicts his experience and all depends on the credibility of the witnesses—their number, integrity, and skill—and the existence of contrary testimonies. There was only one witness, the ambassador, and so the king concluded he was a liar.

Throughout history and on to the present day, telling fact from fake has remained a challenge. Even if we know by hindsight, we often cannot know by foresight. Consider another historical example. In the year 1515, the German master painter Albrecht Dürer carved a woodcut of a rhinoceros (figure 11.1, top). Dürer had never seen this animal with his own eyes; he relied on a sketch and a newsletter sent by a merchant from Lisbon. Similarly, relying on the testimony of others, the English cleric Edward Topsell carved a woodcut depicting a unicorn (figure 11.1, bottom).[2]

Imagine you were born in Dürer's hometown of Nuremberg in 1616, the year the Roman Inquisition demanded that Galileo abandon his belief that the earth and planets revolved around the sun. You grew up in a time when facts were decided by religion rather than science and traveling far from home was strenuous. Thanks to the invention of the printing press, you were able to see these two woodcuts. Would you have believed such animals really exist? A huge animal with a horn on its nose and a massive protective armor that surpasses the heaviest battle horse? Or this slender horse-like animal with a very long, thin horn? Unicorns had at least been painted before, most famously by Leonardo da Vinci, and its long horns had been found and exhibited (these were in fact narwhal horns, but people then had never seen a narwhal either). Your conclusion might have been that both animals are equally plausible or implausible.

The cases of frozen water, the rhinoceros, and the unicorn illustrate how difficult it can be to tell fact from fiction. After the printing press was invented, Europe witnessed a wave of fake news on printed broadsides and pamphlets. Each revolution in communication technology, from the printing press to the internet, has opened new gates for a flood of disinformation. One might object that modern technology has made it easier to sort out the fakes from the real thing. If photography had then existed, the ambassador from Holland might have convinced the King of Siam by showing him a photo of people skating on hard water. Yet photos themselves provide no failsafe proof. Already in 1917, two Yorkshire girls aged ten and sixteen took photographs of themselves with tiny fairies dancing around their heads.[3] Arthur Conan Doyle, the creator of Sherlock Holmes and, unlike his famous character, a staunch believer in spiritualism, was

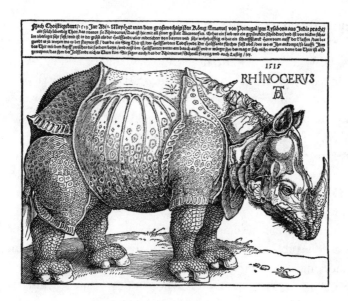

Of the *UNICORN*.

Figure 11.1
Which one is a real animal? Albrecht Dürer's woodcut of a rhinoceros from 1515 and Edward Topsell's woodcut of a unicorn from 1607. Source: Wikicommons. Unicorn: https://commons.wikimedia.org/wiki/File:Oftheunicorn.jpg; Rhinoceros: https://commons.wikimedia.org/wiki/File:D%C3%BCrer_rhino_full.png.

convinced that these photos were genuine. Speculation is still rife about how the girls got the fairies into the pictures; they never disclosed their technique. With today's software technology, photos and videos can easily be edited by everyone. In a deepfake, a person in an image or video is replaced with someone else's face or body; this can already be done live, such as during a Zoom meeting. Or what politicians say in a video can be altered, literally putting words into their mouth. In fact, as technology advances, it may grow increasingly difficult to separate truth from edited truth. That also means that people need to become smarter in understanding who is behind a message, photo, or video and in evaluating the trustworthiness of content.

Fact-checking has become a buzzword, and currently more than 100 fact-checking projects around the world are in operation. International Fact-Checking Day appropriately takes place straight after April Fool's Day, on April 2. Yet facts alone are not synonymous with the truth; the facts can be 100 percent right, but the story can still be misleading, if driven by a hidden agenda. And who checks the fact-checkers? For instance, when the now-defunct conservative magazine the *Weekly Standard* became an approved fact-checker at Facebook, progressive watchdog groups protested, having previously criticized the *Weekly Standard* for pushing false claims about Obamacare.[4] What we have today are some fact-checking organizations accusing others of being partisan fact-checkers.

Fakes

In George Orwell's *1984*, Winston Smith is an employee of the Ministry of Truth, whose mission is to propagate lies. Winston's greatest pleasure in life is his work, which entails altering unkept promises and false predictions made by Big Brother to fit the current facts. When the Ministry of Plenty, which maintains a regime of scarcity and starvation, reduced the chocolate ration after promising that it would not be lowered, Winston converted the promise into a warning that there might be a reduction. When the Ministry of Peace, which engages in warfare, predicted no invasion on the South Indian front by Eurasian enemies but then it happened, that paragraph in Big Brother's speech was turned into a prediction of the invasion. Minute by minute, the past is updated in newspapers, books, posters, films, soundtracks, and photographs. Big Brother is never wrong.

In the pre-internet setting of Orwell's book, the Ministry of Truth employs thousands of employees to collect all copies of newspapers or books, destroy them, and replace them with the altered versions. Today, digital technology enables every individual to alter evidence, photos, and videos in a way that the forgery is hardly detectable. No Ministry of Truth is needed. As a consequence, the distinction between what is real and what is fake blurs. And that is not because people have to trust the testimony of others when they have not experienced an event themselves, as John Locke thought. Even after seeing a picture or video on the internet, we can no longer be sure to what degree it represents reality. Editing pictures of oneself has become normal conduct. Social media overflows with photos of beautifully optimized faces and bodies that may make those with a less spectacular physiognomy feel inferior, and in turn induce them to optimize their own photos. Manipulation has become so normal that unedited photos are often specially marked in social media as "no filter." When adolescent girls aged fourteen to eighteen were shown pictures of peers in which body parts appeared artificially slender, they found these pictures more "natural" than unedited ones—even though the girls were perfectly aware that the pictures had been manipulated.[5] When the meaning of a term changes, thinking changes. Artificial is Natural.

People spread fake pictures or fake news for mundane reasons such as social comparison and getting attention and likes. But they also do so to exclude and punish social groups. These motives are political and nothing new. When the black death hit Europe in the mid-fourteenth century, Jews were accused of deliberately poisoning the wells and were massacred in Toulon, Barcelona, Basel, and the rest of Europe. In Strasbourg, they were burned alive as a preventive measure before the plague had even arrived. Jews were also accused of slaughtering Christian children and using their blood for religious rituals, a myth renewed by QAnon and other twenty-first-century conspiracy communities.[6] The COVID-19 pandemic provided new fertile ground for blaming other cultures. At the beginning of the pandemic, citizens of Asian descent were attacked in many non-Asian countries, and Chinese restaurants reported a drop of 50 percent in business as customers shunned them. Rumors and stigmas spread rapidly through social media, with claims that Chinese cookies, rice, and Red Bull were contaminated with the virus, and that the whole COVID-19 pandemic was a conspiracy against Trump to drag him down in the presidential election.[7]

Conspiracy theories have led to lower acceptance of masks, social distancing, and vaccines during the pandemic, and have had a disastrous impact on HIV prevention, Zika, and Ebola in the past.

I Have Said It Thrice

Before the Brexit referendum, Boris Johnson explained to the British how they had become the victims of the European Union, a bureaucratic monster that chains free markets and suffocates its citizens with ludicrous rules. One of the silly rules, he wrote in the *Telegraph*, is that children under eight aren't allowed to blow up balloons. Another one is that you can't recycle a tea bag.[8] Months before the Brexit vote, when questioned during a parliamentary Treasure Committee hearing as to which EU regulations actually stated this, Johnson bluntly told the committee chairman that it says so on the website of the European Commission. The chairman, however, had the toy safety requirements in front of him, and there was no such rule—no ban on blowing up balloons, not even a requirement that parents should be present. The safety requirement merely asked for warning labels on the packaging that children under eight can choke on balloons. Johnson had fabricated the story. When the chairman also pointed out that there was no EU legislation that forbids recycling tea bags, Johnson had to admit that this decision was in fact taken by the local Cardiff City Council. But what about the coffins? Johnson had also told the public about ridiculous European legislation governing the weight, dimension, and composition of coffins, another myth. Or prawn (shrimp) cocktail chips? According to Johnson, one of the biggest threats to the freedom of the British food industry was the EU ban on prawn cocktail-flavored potato chips. He told the tale of bravely traveling to Brussels to confront the bossy female bureaucrat responsible for outlawing them, a symbolic humiliation of British democracy. In his own words, "We will die in the last ditch to preserve the prawn cocktail flavour crisp." Once again, no such EU ruling existed, and probably also no such woman, as he had made up many of his stories about Brussels.[9] Yet in the end, that did not matter. The fairy tales paid off.

Why would people believe in fake news? The common answer is that people are stupid, not interested in evidence, and want to have their opinions confirmed. That may be true in some cases, but there is a more interesting explanation based on a fundamental psychological law:

Reiteration effect: The more frequently an assertion is repeated, the more believable it becomes, independent of whether it is true or false.

In Lewis Carroll's poem *The Hunting of the Snark*, the Bellman proclaims: "I have said it thrice: What I tell you three times is true."[10] The reiteration effect works with political news that confirm our prejudices but surprisingly also with neutral news and trivia. I discovered this myself in an experiment I conducted where participants were given assertions to read such as "The People's Republic of China was founded in 1947" and "There are more Roman Catholics in the world than Muslims."[11] For each assertion, participants had to express their confidence in whether it was true. Two weeks later, they were given a new set of assertions containing some of the old ones. The same procedure was repeated another two weeks later. Participants' belief in the repeated assertions increased with every repetition, whether they were true or not.

Note that the reiteration effect assumes lack of knowledge. Someone who knows for sure that the People's Republic of China was proclaimed in 1949 by Mao Zedong after the Chinese Civil War ended will not be influenced by repetitions of a false fact. In general, the larger people's ignorance about a topic, the more likely they will be subject to the reiteration effect. Predictions about the future are always uncertain, and here reiteration has a fertile field for persuasion—and for influencing what will happen. The Roman statesman Cato is said to have reiterated his call to destroy Carthage at the end of every one of his speeches until his call became reality. The Russian revolutionary Vladimir Lenin, among others, is reported to have said that a lie told often enough becomes a truth. In social media, a fake news cascade begins with a rumor and continues when others propagate the rumor by retweeting it. As a consequence, people may hear the same message repeated several times, and the reiteration effect kicks in. Each time the news sounds more believable, until we firmly believe it.

Yet repetition alone is not the entire story. Five repetitions are more powerful if they come from five different people than from the same person. Some of these five people may even be social bots. These algorithms present themselves as people in order to influence human users on social media. Coordinated networks of bots, called botnets, can be effective in spreading messages to real users. In one study, a botnet spread Twitter hashtags, such as #getyourflushot to encourage users to vaccinate and #turkeyface for Photoshopping a celebrity's face onto a turkey.[12] Within a short time, 25,000 human users followed the botnet, most of them following several bots. Human users were influenced by how many different bots repeated a message, not just the number of times a message was repeated. In this way, the spread of news through social media differs from the spread of a virus in infectious diseases. While the chance of getting infected increases with each exposure to the virus, no matter whether it is from the same or a different infected person, the chances that information spreads through Twitter depend crucially on both the number of repetitions and the number of different people or bots that repeat the content.

When we talk about fake news, we tend to think of politicians' lies or far-fetched conspiracy stories such as that SARS-CoV-2 was genetically engineered by a Chinese biological weapons program, caused by the electromagnetic fields of the 5G mobile networks, or was manufactured by Jews to cause a global collapse of the stock market and profit from insider trading. Yet we hardly notice the mass of "normal" fake news. It is much subtler than plainly wrong facts and sometimes not even deliberate but inadvertently dreamed up.

Blunders

One source of unintentional fake news are blunders. A blunder is due to someone making a mistake because of neglecting to think or read a source carefully. They may not be deliberate fakes, but they can be equally misleading. Often, a reader can immediately see that some news must be a blunder, even if they'd like the story to be true. If you enjoy jogging, for example, you might be receptive to this media headline about a study on longevity: "For every hour of daily jogging, you live 7 hours longer."[13]

To invest one hour and gain seven—what more can you ask for? As a passionate jogger, you can start calculating how much longer you will live than the couch potatoes in your neighborhood. But pause to think for a moment. If that claim were true, we could literally run ourselves into immortality. Jogging, say, four hours a day would mean living twenty-eight hours longer. That is more than a twenty-four-hour day, and thus our life expectancy would increase each day. Clearly something is wrong here. Indeed, the original study made no such claim. What it said was that this effect holds for two hours of jogging per *week*, not for more. More precisely, the seven-hour figure was estimated this way: a group of forty-four-year-old joggers who run for two hours per week spend a total of 0.43 years running by the age of eighty and win 2.8 years of increased life expectancy. These are the numbers that were translated into the headline proclaiming one hour of running for seven hours of living longer. More running is not necessarily better. On the contrary, excessive running can increase the risk of heart disease and shorten one's life.

The jogging headline is easily shown to be false—a bit of thinking suffices. Alternatively, one can switch from the website that featured the headline to more trustworthy sources on scientific studies, such as the *New York Times*, where one can find the correct claim that the gain in life expectancy caps at around three years.[14]

Fairy Tales about Algorithms

Much of what you hear about AI is also riddled with fake news, often subtle, preying on awe or fear to capture your attention or to sell a product. The entire history of AI has been soaked with overinflated promises and hopes, not so unlike those in stock markets that lead to bursting bubbles. The big downs in this roller-coaster ride were the "AI winters" of the 1970s and 1980s, with cycles of enthusiasm and hype followed by disillusionment and funding cuts. As a result, AI became a dirty word for quite a time. For instance, as recently as in 2011, when the supercomputer Watson was created, IBM shied away from calling it AI and instead dubbed it "cognitive computing," fearing that otherwise no one would take Watson seriously. Since then, we have seen true advances in deep neural networks and computing power but also unqualified, glorified claims about the general superiority of technologies over humans. Many of these tall tales are motivated by making profit, getting funding, or wishful thinking. Promises about the future are cheap but hard to evaluate. One can always say, "If it's not now, it will be soon."

There is another, more subtle and interesting form of hype. It works by using terms the audience likely misunderstands and by rewriting the past. This technique recalls Winston Smith's attempts at the Ministry of Truth to alter reality to create the impression of an omniscient Big Brother. I will briefly describe only a few cases, but the general message is to beware of extravagant claims. The first uses a method of persuasion that has stood the test of time in centuries of advertisement: to describe a product by using a term that that suggests properties it doesn't have.

Full Autonomy

As we saw in chapter 4, the Society of Automotive Engineers distinguishes five levels of automation in cars, from cruise control (Level 1) to self-driving cars (Level 5) that drive safely everywhere and in all traffic conditions without a human actively monitoring their operations. These levels are not the same. Nevertheless, car manufacturers often call their Level 2 or Level 3 cars "self-driving" in advertisements, which tend to be repeated in the media. Take Tesla, whose marketing claim since 2016 has been that "all Tesla cars being produced now have full self-driving hardware"[15] and whose owner Elon Musk promised "full autonomy" by 2018.[16] At the same time, the advertisements correctly add that the system needs active monitoring by the driver, or that the proper software has yet to be developed. The expression *full self-driving hardware* means that the computers can steer, brake, and accelerate, but the main problem is having the intelligent software. By reiterating the terms *full autonomy* and *self-driving cars*, more and more people have begun to believe that these vehicles are actually on the roads. This confusing and suggestive language has been noticed and has led to lawsuits. For instance, the German Center for Combating Unfair Competition has sued Tesla for making false promises to consumers.[17] The center, it should be mentioned, represents more than 1,000 companies, including Tesla's competitors Audi, BMW, Daimler, and Volkswagen.

Why don't companies advertise the amazing progress in Level 2 driving that might eventually lead to full Level 3 driving? Car engineers can be proud about this new technology. Honesty and modesty might be the wiser option. It could avoid another downfall in reputation the moment people begin to notice that the words do not mean what they are led to believe.

Curing Cancer

A favorite turf of AI hype is health care. A case in point is IBM's marketing of Watson for Oncology, which, as chapter 2 explains, has little to do with what the computer is actually able to do. In the words of Peter Greulich, a former IBM manager, "IBM ought to quit trying to cure cancer. They turned the marketing engine loose without controlling how to build and construct a product."[18] Commercial companies have vested interests in selling their products, and IBM's marketing team has been quite successful in creating the illusion that Watson can cure cancer. They have been so successful that authors of popular books on AI have become company salespersons.

A second classic rule of persuasion is to present the reader with a false set of choices. You are told that the most important medical decisions are increasingly based on computers such as IBM Watson, and that these know you better than you and your doctor can. The only downside is that the AI will have to know everything about you and will decide what is best for you. You can't decide anymore about what you eat and what treatment you get because that's the very point. Here is your choice:

- Maintain your privacy and free decision in health care *or*
- Get access to far superior health care.

The logical conclusion, we are told, is that most people will choose superior health care and surrender to AI authority.[19] That may well be, but the choice is a wrong one. It reminds me of financial advertisements that ask us to choose between managing our own money badly or handing it over to a portfolio manager who will make far superior investments. That choice is wrong because studies have found no evidence that, on average, portfolio managers will increase your wealth more than you would on your own.[20] It's an advertising ploy, aimed at increasing the managers' own wealth. Just as the financial ad discourages you from taking responsibility for your own wealth, authors who pose medical choices like the one above discourage you from making your own informed choice—and, to add insult to injury, discourage young people from considering studying general medicine.

As I argued in chapter 2, we can reap the benefits of AI to patients' health only if the two existing chronic diseases of health care systems—conflicting interests and risk illiteracy—are faced head-on and cured. Otherwise, AI will be gamed and able to contribute little to patients' health. A supercomputer alone does not help patients. AI is always AI designed and marketed by people. To believe that mere computing power and smart algorithms by themselves will generate superior health care is an illusion. It means failing to understand that diseases are more complex and health care systems are more dysfunctional than many computer scientists anticipate.

So far, we have seen two classic methods of persuasion, using deceptive terms and presenting wrong choices. Another method is to use a well-known success story and refashion it so that it appears to support one's favorite message: that algorithms lead to superior decisions.

Moneyball

Moneyball, the 2003 US bestselling book by Michael Lewis that spawned a movie starring Brad Pitt, tells the story of Oakland A's general manager Billy Beane and how he led his baseball team to greatness on a shoestring budget. The second hero is Bill James, a statistician, and their shared insight is that baseball can in essence be reduced to data and algorithms, which they call *sabermetrics*. According to Lewis, for more than a century, managers and general managers relied on their guts or educated guesses for spotting future major league baseball players.[21] Beane, in contrast, introduced an algorithm to spot "sleepers," that is, unknown or underrated talents: "It is simply a matter of figuring out the odds, and exploiting the laws of probability."[22] In Lewis's account, the players recruited by algorithms were decisive for the Oakland A's success in the early 2000s.[23] In baseball, there is a long debate between relying on the intuition of experienced scouts for picking players and on statistical number crunching.[24] *Moneyball* argues that the revolutionary use of baseball statistics changed the game, one of the great victories of algorithms over expert intuition.

Hollywood is famous for stretching the truth into melodramatic stories. Baseball experts noted that Lewis likewise did not let the facts stand in the way of a good story. Among those facts, the relevant one here is that players selected by algorithms and featured by Lewis actually played relatively little part in Oakland's success. The team thrived primarily because of three superb pitchers known as the Big Three, all of whom were discovered by traditional scouting methods based on intuition and judgment, not by algorithms:

> At the heart of the pitching staff were three dominant starters: Mark Mulder, Tim Hudson, and Barry Zito. All three were early-round picks, highly scouted, and well regarded—Mulder and Zito were selected in the top ten of their respective drafts. This was hardly a case of Beane's spotting sleepers . . . because of nuanced numbers. Indeed, Michael Lewis does not suggest that sabermetrics had anything to do with Beane drafting the three studs who led Oakland to greatness. Indeed, he virtually ignores them. Lewis devoted a few paragraphs to the Big Three (making the strained claim that Beane appreciated them for quirky reasons), quickly dropping them and transitioning to an entire chapter on . . . Chad Bradford.[25]

Bradford was a pretty good relief pitcher, but he performed nowhere close to the Big Three in innings, wins, or savings.[26] In fact, after Oakland lost its Big Three pitchers, the team's successful run between 1998 and 2003 plopped. Lewis's account of the supreme wisdom of algorithms is a well-told story. But it is a fiction, conveniently leaving out the facts that do not fit its narrative. Nevertheless, it has become a staple example used by gifted storytellers and popular writers who want to convince us that AI will soon replace human judgment in all domains, even in ill-defined ones such as spotting future top baseball players.[27] The better alternative would be to ask how baseball statistics and expert intuition can be combined to make better decisions, but that would be a difficult question and not the fabric for a grand tale of heroic success.

The story about the Oakland A's success illustrates a genre that twists the evidence to create the impression that algorithms have been the key to better decisions in situations where they play little role. Such hype helps to sell popular books, but does not help us understand the actual potential and limits of AI.

Personalized Ads: A Bubble about to Burst?

Google earns about 80 percent of its revenue from ads, and Facebook even 97 percent. The advertisers pay these astronomical sums. Therefore, one might expect them to have carefully checked whether the return on their ads justifies the costs. While at Google, Eric Schmidt assured advertisers that they pay only for what works: "Our business is highly measurable. We know that if you spend X dollars on ads, you'll get Y dollars in revenues."[28] Yet there is increasing evidence that this is not the case; rather, it seems unclear in many cases whether personalized ads actually pay.

Every time you search for information about a product online, manufacturers and retailers can bid for paid ("sponsored") search ads that are placed on the top of the unpaid ("organic") search results. Which company wins the bid is determined by an automatic auction. Google and others earn money by every click on the sponsored ad, but not if you click on the organic result. In addition, when you read online content such as sports, advertisers can bid to have display ads shown on the pages being read. These targeted ads are considerably more expensive than nontargeted advertising. They require that the platforms collect as much data about you as possible to predict your clicks. This collection and analysis of personal data is the heart of surveillance capitalism. But there are now a number of reasons to suspect that this system fails to deliver what has been asserted, and instead resembles a bubble that may burst.

Do Sponsored Ads Pay?

When Steve Tadelis, a professor of economics at the University of California, Berkeley, spent a year at eBay, its marketing consultants spoke of how profitable their ad campaigns were. The most successful method was *brand keyword advertising*, he was told. If a search includes a brand keyword, such as "eBay motorcycle," Google, Bing, and other platforms offer to place a paid link to the brand, here eBay, at the top of the organic search results. The consultants asserted that eBay earns $12 for each dollar spent on brand keyword advertising.[29]

Together with two economists at eBay, Tadelis conducted a number of experiments to measure the actual returns. In each experiment, eBay halted brand keyword advertising (on Google, Bing, and Yahoo!) in one set of cities while continuing advertising in others. If the sponsored ads actually worked, eBay's earnings should decline in the periods when the ads were removed. But it didn't. Nor did earnings drop when eBay halted bidding for nonbrand keywords, such as "cell phone" and "used gibson les paul." Only new and infrequent users were influenced by ads, not more frequent users. Tadelis and his co-researchers calculated that eBay by no means earned $12 per dollar spent, as was claimed, but instead lost 63 cents for each dollar spent.[30] Having learned that they were actually losing money on these ads, eBay struck brand keyword advertising from its marketing budget.

Unlike the economists, the marketing consultants were looking at correlations, not causes. Say you run a well-known coffeehouse and hire two people, Jack and Joe, to hand out coupons to attract customers. Soon, half of the customers arrive with coupons distributed by Jack, but few with those distributed by Joe. You might well conclude that Jack's marketing strategy or charisma is superior and has led to some 50 percent of sales. Yet, unlike Joe, who went downtown to give away coupons, Jack stood by the coffeehouse and gave coupons to those people in front of it.

Just as most of the people who arrived with Jack's coupons in hand would have visited the coffeehouse anyway, 99.5 percent of the users ended up on eBay's website without the paid link. All they needed was to click on the organic link to eBay or to go directly to eBay's website. Note that eBay has to pay Google only if you click on the sponsored link, not if you click on the organic link, which is typically just below it.

It is likely that other well-known brands would obtain similar results, but that remains conjecture because experiments on the topic are rarely conducted. An experiment with a less well-known brand, Edmunds.com, a source for automotive information, reported that when brand keyword advertisement was shut off, only half of its normal traffic flowed through the organic search link. The other half likely landed on websites of competitors, who sneakily bid on the keyword "Edmunds."[31] Yet, unlike the eBay experiment, this study could not provide precise measurements of the returns over investment. Thus, companies with low brand recognition might profit from brand keyword advertising, if only to protect themselves from competitors who might otherwise draw away their customers. But that is like being forced to distribute coupons in front of your coffeehouse because otherwise your competitors will stand there and distribute their own coupons.

The general lesson in face of the eBay experiments is that companies should run their own experiments to find out whether their ads actually pay. But even among those firms whose branded keyword ads were not regularly purchased by competitors (large firms like eBay), only one out of every ten stopped brand keyword advertising, and mostly without doing an experimental study. The majority of firms just continued conducting business as usual.

You might wonder why. It is in the best interest of a company to know how effective their advertising is, but that does not necessarily apply to its marketing department. The department can secure a larger budget and more personnel if their campaigns are seen as brilliant. It also competes with the print and TV marketing group, which provides another internal conflict of interest. Besides conflicts of interest, the prevailing reliance on correlations rather than experiments leads to overblown estimates of campaign effects.[32] For instance, Google shows its customers how to calculate the return over investment in a way that leads to these inflated estimates.[33]

Do Display Ads Pay?
In an analysis of twenty-five large-scale digital advertising field experiments from well-known retailers and financial service companies, researchers

from Google and Microsoft concluded that it is nearly impossible to measure the returns of advertising.[34] Contrary to Eric Schmidt's claim, if the ads have an effect, these are often so tiny that it is difficult to prove their existence. Similarly, an experiment with 1.5 million customers of a nationwide US retailer showed that displaying ads for a clothing line resulted in only tiny and insignificant effects on increased purchases, with one exception.[35] For customers age sixty-five and older, sales increased by 20 percent due to advertising, but almost all of these customers went to brick-and-mortar stores as opposed to ordering online. Two follow-up experiments didn't find an overall increase in purchases either.

Adding to the uncertainty is the likelihood that display ads are no longer as effective as they might once have been. First, users today pay less attention to ads. Since 1994, for instance, when the first banner ads were displayed, the click-through rate has plunged from a remarkable 44 percent to 0.46 percent in 2018.[36] Second, many consumers are annoyed by the steady flood of ads and increasingly rely on ad blockers. And third, advertisers face click fraud. Entire bot nets and human "click farms" are employed to deliver click-throughs on ads in order to make campaigns look better than they really are. These services are even hired by crafty advertisers to click on the ads of their competitors in order to mislead them about the effect of their ads and increase their costs, thereby depleting their budget. Studies estimated that more than a quarter of website traffic showed nonhuman signals and that more than half of all ad dollars spent on display ads were lost to fraud.[37] The upshot is that advertising platforms profit from ad fraud, including from every false click.

All in all, there is considerable uncertainty about whether and when sponsored and display ads increase sales, and if they do, whether such an increase justifies the increased costs. Diminishing attention, ad blocking, and fraud tilt the scale against the widely claimed great benefits for advertiser. This recalls another situation years before the financial crisis of 2008, where the big global rating agencies gave AAA ratings to banks' toxic mortgages, which allowed banks to sell these for exaggerated prices. The rating agencies were not neutral; the banks paid them for the perfect grades, just as advertising agencies profit from overstating the value of ads. The high prices for ads were one of the reasons why eBay began to check what they were actually getting for their money.[38]

If a sufficiently large number of companies followed eBay's lead, rose above their internal conflicts of interest, and conducted their own experiments, the advertising bubble might well follow the fate of the pre-2008 financial bubble. That would not only benefit the advertisers' budgets but also have the potential to change society at large. For one, it would free the brainpower of bright young researchers who currently focus on predicting clicks, so that they can apply their talents to something more useful.

Crucially, if further experiments confirm that many ad campaigns do not deliver the promised return on investment, more companies might reconsider spending huge amounts on such personalized ads with little return. The pay-with-your-data model would crumble, and with it, the desire of tech-companies to sell our attention and time. That would bring about the end of ad-based surveillance capitalism as we know it.

Checking Trustworthiness

Checking the trustworthiness of sources entails more than checking facts. It requires digging out who is behind the information we get, what the underlying intentions are, and whether the information is correct. Traditionally, investigative journalists have been the gatekeepers of trustworthy news. But their number has been dwindling since the rise of large media conglomerates in the 1980s, for whom publishing the truth about corruption or corporate crime may not be in their best financial interest. Advertisers have reduced spending with media that reported unfavorable details about their business practices. The rise of social media further contributes to weakening traditional media's loss of control, with positive and negative consequences. On the one hand, people can report on issues such as corruption and violations of human rights not covered by media conglomerates. On the other hand, everyone can easily spread rumor, lies, and hoaxes. Now, all of us are called to be our own investigative journalists. How many of us are ready?

Digital Natives

It is widely believed that being fluent on social media is equivalent to being able to navigate through the tricks and traps of the internet. Digital natives know how to keep multiple information at their fingertips, fluidly switching between TikTok and Snapchat while texting a friend and uploading a selfie to Instagram. With hours of online experience every day, so one might think, skills to judge the credibility of information evolve naturally.

To see how skilled at this digital natives actually are, Stanford researchers assessed some 900 students from middle schools, high schools, and colleges in twelve US states.[39] They asked the students to evaluate online sources via questions such as "What is the evidence?" and "Who is behind the information?" For instance, middle schoolers read an online article titled "Do Millennials Have Good Money Habits?" written by a bank executive and sponsored by Bank of America, which argued that many millennials need help with financial planning. Then they were asked to think about one reason why they might not trust the article. Surprisingly, most students did not consider authorship or sponsorship and the resulting conflict of interest as grounds for skepticism. In another test, they were asked to look up the home page of the online magazine *Slate* and determine whether content was an advertisement or a news article. The middle schoolers could easily identify traditional ads with coupon codes. But over 80 percent believed that a native (paid) ad clearly identified with the words "sponsored content" was a real news story.

Are high school students better at evaluating online content? In that study, they were asked to evaluate two Facebook posts announcing Donald Trump's candidacy for president, one from Fox News and one from a similar-looking fake account. The real post contained a blue checkmark indicating that the account had been verified as legitimate by Facebook. Only a quarter of the high school students were aware of the significance of the blue checkmark, and one out of three found the fake Fox News account more trustworthy.

Another post shown to the high school students, from the photo-sharing website Imgur, featured a picture of malformed daisies along with the claim that the flowers had nuclear birth defects following Japan's nuclear disaster at Fukushima (figure 11.2). The question posed by the experimenters was whether the picture provides strong evidence about the conditions near the nuclear power plant. A naïve person would be taken in by the compelling picture, while a critical reader would be more careful and note that there is no proof that the picture was actually taken near Fukushima. As one critical student noted:

Figure 11.2
Nuclear birth defect? Mutated flowers, similar to those in one of the tasks of the Civic Online Reasoning Test. Source: Perduejn/Wikimedia Commons, https://commons .wikimedia.org/wiki/File:MulesEarFasciated_107393.jpg.

> No, it does not really provide strong evidence. A photo posted by a stranger online has little credibility. This photo could very easily be photoshopped or stolen from another completely different source; we have no idea given this information, which makes it an unreliable source.

But only a minority of the high school students pointed out the unknown source and the lack of evidence. In contrast, three-quarters of them did not question the source or the evidence at all. Their reasoning was in line with the argument of one student:

> This post does provide strong evidence because it shows how the small and beautiful things were affected greatly, that they look and grow completely different than they are supposed to. Additionally, it suggests that such as disaster could happen to humans.[40]

All in all, the far majority of high school children had never learned to reason critically about posts—despite being digital natives.

What about college students? They were asked to evaluate whether a particular website was a reliable source of information about the minimum wage and were directed to an article on minimumwage.com titled

 **Minimum Wage**
FACTS & ANALYSIS   ALL STATES   MINIMUM WAGE 101   RESEARCH   BLOG   ABOUT   ADS

Denmark's Dollar Forty-One Menu

Thursday, **October 30, 2014**, 9:00 am

Proponents of raising the minimum wage often point to Scandinavian countries like
Denmark as models for American labor policy. But the devil is in the details. Take
this week's *New York Times profile* of the comparatively high Danish minimum wage,
for example. The authors ask, if the Danes can do it, why can't the United States?

In the midst of a mostly-fawning piece on Danish labor policy, the authors
unwittingly answer their own question: It would lead to higher prices and fewer job
opportunities.

The piece points out that the associated higher labor costs mean that a Big Mac in
Denmark costs 17 percent more than in the United States – $5.60 versus $4.80.
Other analyses put the price discrepancy at around double this. For example, the
equivalent of the "Dollar Menu" in Denmark is $1.41, and an extra value meal is
nearly 40 percent more.

OCTOBER 30, 2014

**Bernie's $15 Plan Will Cost
Georgia 106k Jobs**

OCTOBER 30, 2014

**No Blue Wave is a Good Sign
for Minimum Wage**

OCTOBER 30, 2014

**Fact-Checking Biden on
Minimum Wage**

Figure 11.3
Is this a reliable source of information about minimum wage? One of the tasks of
the Civic Online Reasoning Test. Source: https://www.minimumwage.com/2014/10
/denmarks-dollar-forty-one-menu/.

"Denmark's Dollar Forty-One Menu" (figure 11.3). The article takes up a
question raised by the *New York Times*: If Denmark can pay its workers a
comparatively high minimum wage, why can't the United States? It argues
against raising the minimum wage because that would increase labor costs
and make Big Macs cost more. The equivalent of the "Dollar Menu" in
Denmark is $1.41, making Danish fast food restaurants far less profitable.
Therefore, minimum wage would raise prices in the US and eliminate hundreds of thousands of jobs.

The website looks reliable and describes itself on the "About Us" page as
a "a project of the Employment Policies Institute (EPI)," which is "a non-
profit research organization dedicated to studying public policy issues surrounding employment growth."[41] In response to the article, the *New York*

Times reported that EPI "is led by the advertising and public relations executive Richard B. Berman, who has made millions of dollars in Washington by taking up the causes of corporate America." A reporter who visited the EPI headquarters found that nobody is employed there; it is just one of Berman's many online entities. That information, of course, cannot be found on minimumwage.com. In order to find out, the students would have to exit the website and search for what is known about its sources—a process that is known as *lateral reading* (as opposed to *vertical reading*, that is, how we read printed text). Yet, even though that possibility was explicitly mentioned in the instructions, the vast majority of students never ventured beyond the site. Instead, they trusted its appearance and what was said on the "About Us" page. As one student explained:

> I read the "About Us" page for minimumwage.com and also for Employment Policies Institute. EPI sponsors minimumwage.com and is a nonprofit research organization dedicated to studying policy issues surrounding employment, and it funds "nonpartisan" studies by economists around the nation. The fact that the organization is a non-profit, that it sponsors nonpartisan studies, and that it contains both pros and cons of raising the minimum wage on its website, makes me trust this source.

This student reasoned soundly but based solely on how the organization portrayed itself. Less than 10 percent of college and high school students went beyond the surface appearance of the web page and were able to critically evaluate the site.

Regardless of what one thinks about minimum wage, tracking down who is behind a site reveals its hidden agenda. From middle school to college, however, most students rarely asked who stood behind an online source, did not consider the evidence for the claim presented, and did not consult independent sources to verify the claims. Rather, they took what was said at face value and were captivated by vivid photos and graphic design. Even when encouraged to do an internet search, most did not move beyond the original website. All in all, they were easily duped. Being a digital native does not mean being digitally savvy.

Professionals and Elite Students

Surely professionals and students from the best universities are better skilled at evaluating trustworthiness and evidence? To answer this question, two of

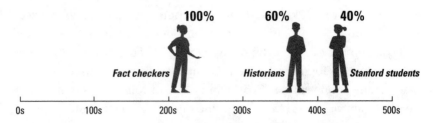

Figure 11.4
Average time that professional fact-checkers, historians, and digital natives (Stanford undergraduates) needed to find out who was behind the minimumwage.com website and the Employment Policies Institute. The percentage of members of each group who found the answers is shown on top of the icons for each group, and the average time it took them is shown in seconds (see text). See Wineburg and McGrew, "Lateral Reading."

the researchers from the previous experiment rounded up ten professional fact-checkers employed at well-regarded news and political fact-checking organizations, whose job is to ascertain truth in digital media, and ten professors of history, whose daily bread is to evaluate the credibility of written texts and the circumstances of their creation.[42] They also enlisted twenty-five undergraduates at Stanford, one of the most competitive universities in the world located in the heart of Silicon Valley. These students represent our digital future.

Each of the fact-checkers, historians, and students was given eight minutes per site to evaluate minimumwage.com and two other websites. All of the fact-checkers—but only 60 percent of the historians and 40 percent of the students—found out which sponsors were behind minimumwage.com and the Employment Policies Institute. Moreover, the fact-checkers were much speedier at doing so, on average within 205 seconds, while those students who also found out took twice as long (419 seconds), with historians in between (361 seconds; figure 11.4).

What was the fact-checkers' secret? Consider the fastest fact-checker. She looked at "Denmark's Dollar Forty-One Menu" for only six seconds before going to the "About" tab, where she learned that the site was a project of the Employment Policies Institute, and then opened a link to it in a new tab alongside minimumwage.com. After spending just three seconds on the EPI home page, she clicked on its "About" and read that it was founded in 1991 and is "a non-profit research organization dedicated to public policy issues." "This is profoundly not helpful," she remarked, and then googled the EPI. Reading the snippets, she skipped the first results, selected Source-Watch's entry on EPI, and quipped: "So this says it's one of several front groups created by a PR firm." She scrolled until finding a linked quote from the *New York Times* reporter who had tried to visit the EPI office and found no evidence that it even existed. She clicked on the citation, provided by SourceWatch, and checked the claims by going to the website of National Public Radio, which provided a transcript of an interview with the reporter. Within just two minutes, she discovered that minimumwage.com and the EPI were not what they appeared to be:

> Obviously, this isn't a legitimate organization, based on the reporting of this *New York Times* reporter. He talks about actually going there, he doesn't see any evidence at all that they actually had an office, there are no employees, all the staff there actually work for the PR firm.[43]

Why did all fact-checkers but not all of the students and historians detect the hidden agenda of the website? The answer is that many students and historians used different strategies; in particular, they didn't budge from the landing page and read through it in its entirety. If they did stray from the page, they did so much later in order to click on links that matched their personal interests. All of them no doubt understood that websites are carefully designed and may be financed by groups that promote particular interests, often partisan ones. Yet most lacked the skills to find out. These skills are rules that can be easily learned.

Smart Rules for Judging Trustworthiness

Assume you are contacted by a new website and have limited time for research. The site may ask you to sign up for a movement to decrease minimum wage, increase tobacco tax, legalize marijuana, sign a petition against gay marriage, donate money to their cause, or simply ask you to pass on news. You want to find out whether the site is trustworthy. Here are four rules that professional fact-checkers use:

1. *Lateral reading. Leave the site early, before reading everything, and engage in lateral reading.* Take only a brief glance at the content and then move on to other sites to do a background check on its credentials and agenda.

2. *Exercise click restraint.* When you get search results, do not click on the first entry on the first page. Use snippets to search for clues about the

site's reliability, that is, scan the entire first page (or even more pages) of results in order to make a wise first click.

3. *Go back and read.* Once you know more about the organization behind the site, read the text carefully with an understanding of the agenda.

4. *Ignore surface features.* Don't pay attention to the design of websites or to top-level domains like .com or .org.

A quick way to find out who is behind a website is to first go to "About" on the site and then leave the site to search the organization's name, look up independent sources, and go from there. This step requires some knowledge of sources and organizations, such as how they are aligned with political parties or with societal issues. Click restraint requires some knowledge of online structures, such as that the first result presented on a search engine is not necessarily the most relevant for your search. Results may make the top of the list because of a vocal online group, such as MMR anti-vaxxers, not because of the quality of evidence. Or it is a disguised ad. That first click can be destiny; it may send you on the wrong trail. Always think before you click.

Not-So-Smart Rules

The four rules have been shown to improve students' evaluation of online sources.[44] Yet many digital natives do not seem to be aware of them. Instead, they may have been following old advice, which came in the form of a checklist. Consider this widely disseminated list for evaluating web pages, which can be found on hundreds of websites—including the websites of many universities. It asks you to evaluate a page on the basis of five criteria:

- *Accuracy.* If your page lists the author and institution that published the page and provides a way of contacting him/her and . . .

- *Authority.* If your page lists the author credentials and its domain is preferred (.edu, .gov, .org, or .net), and . . .

- *Objectivity.* If your page provides accurate information with limited advertising and it is objective in presenting the information, and . . .

- *Currency.* If your page is current and updated regularly (as stated on the page) and the links (if any) are also up-to-date, and . . .

- *Coverage.* If you can view the information properly—not limited to fees, browser technology, or software requirement, then . . . *You may have a Web page that could be of value to your research!*[45]

Or you may not. This checklist does *not* ask the user to leave the page and seek independent information. It assumes that all relevant information appears on the page and that investigating it carefully is sufficient. It deals with features and appearances of websites to which companies can always adapt, while fact-checkers' rules deal with critical actions that *we* can take, independent of any modifications made by companies. Versions of this list go back to 1998, the early days of the internet. It can provide clues about reliability, but it is no match for current strategies of persuasion and deception.

In reaction to these kinds of lists, many websites that want to hide their agenda try to fulfill the criteria perfectly. Take, for instance, the minimum-wage.com website, which passes all criteria apart from the question of its domain. It lists authors and institution and provides contact information, which satisfies accuracy. There is no advertising and the information looks objective; thus, it passes the objectivity test. Finally, the site and links are updated, and you can view the information properly without paying a fee, giving a thumbs-up to currency and coverage. Working through this list can provide a false sense of security. The fact-checkers succeeded because they knew that the website designers who want to hide their agenda would be careful to satisfy as many items as possible on this standard pre-2000 list.

Governments pour millions into equipping students with digital tools, from tablets to interactive whiteboards. Equally important, however, are investments in a general curriculum on staying smart in a smart world. Few governments have realized its importance to date. It comes as little surprise that Finland—which ranks consistently as one of the top countries worldwide with respect to happiness, press freedom, social justice, and gender equality, and whose schools consistently excel in international tests of math, science, and language skills—is also at the vanguard of teaching students to tell facts from fakes, pseudoscience from science, and gossip from information. Its best-known digital literacy toolkit, called *Faktabaari* (Fact-Bar), is taught from elementary to high school.[46] Finland also launched the award-winning massive open online course "Elements of AI," available in many languages. Learning in school how to become digitally savvy should not be confined to the land of happy and egalitarian Finns.

The Cement of Society

Trust is the cement of a society. In small villages centuries ago, everyone could closely watch everyone else, and people knew what to expect from whom. Little trust was needed, except, as Martin Luther noted, in God.[47] If someone cheated, stole, or lied, this person could be identified quickly and punished or excluded from the community. Trust became important when human societies grew larger and more mobile.[48] Merchants trading across countries and continents had to trust—without trust, there is no trade.

The invention of the World Wide Web made our social and economic relations even more dependent on trust. More than ever in human history, we interact with strangers whom we may never meet face-to-face. Should we trust a person whom we only know through online dating? Should we trust secret algorithms that calculate who should be considered for a job, a loan, or social welfare? How many of the five-star ratings of a product we consider buying are genuine and how many are bought? Can we trust social media platforms if their business is to sell our attention and time to advertisers? And what about governments? According to a research team at the University of Oxford, some seventy governments worldwide have built social media misinformation teams to spread lies and conceal truths.[49] Should we just shrug and scroll through the news feed to get another dose of dopamine? We can close our eyes and keep going as happy consumers, hoping for the best.

The alternative is to become digitally keen citizens. If homes, factories, and cities become smart, why not people? The digital world has made disinformation cheaper and more scalable than ever before.[50] But it also offers tools to find out about the trustworthiness of people and sources. We can get information about what AI can easily do and what it cannot, and we can consider how the pay-with-your data business model sells our time and attention. And we can vote for politicians who work together with the public to better understand and regulate the benefits and potential harms of AI. The General Data Protection Regulation of the European Union is a first step toward winning back trust. But even more political courage is needed to translate its general ideal of transparency and dignity into concrete measures. So far, many platforms try to outsmart the regulations, for instance, by making "accept all cookies" a single click option while other options require annoyingly extensive scrolling and multiple clicks.

Transparency starts with simple, concrete measures, such as making it easy for people to understand what they actually consent to if they click "I accept." Many information platforms hide the answer in lengthy, sometimes twenty-page terms-of-use agreements in a barely decipherable font, which are not written to be understandable. People are offered a choice between two dishonorable options, either to give uninformed consent or to spend hours trying to comprehend these oversize documents before entering a new platform. This forced choice is an affront to human dignity.

I have argued that there is a simple solution to this state of affairs: to require tech companies to replace unreadable documents with a one-pager that honestly and clearly explains what personal information is extracted and sent to which third parties and explicitly asks your consent to the platform becoming the new owner of your pictures and data.[51] To get there, we need engaged policy makers who are willing to fight for this change. But the fight for human dignity continues at a deeper level. Social media platforms are owned by a few immensely rich men who buy up competitors rather than promote market competition. As we have experienced with banks in the financial crisis, these platforms can pose a threat to a healthy democracy by becoming too powerful and too big to fail. A healthy economy thrives from more competitive innovation and less centralization. Rather than moaning to our friends that our children get too little sleep and are depressed by social media, we need to persuade governments to attack the root of the problem, to stop the surveillance business model.[52]

All these are measures that would allow both tech companies and politicians to win back the trust that is lost as people become increasingly aware of the amount of secrecy and misinformation, and the degree to which we are all subject to commercial and governmental surveillance. We should be able to enjoy social media without thousands of engineers and psychologists constantly working on further addiction methods. We should be able to profit from AI in tasks where it delivers better and faster than humans, without being misled to expect that it can predict all of our behavior and improve all aspects of our lives.

The original dream of the internet was to open the door to the information age. Now we find ourselves in a world that is both an information and a disinformation age. The latter is a serious threat to human evolution: it can erode trust in institutions we have built for the better of all of us, such as government, science, investigative journalism, and the justice system. We need to fix the internet. Eliminate surveillance business models. Resuscitate privacy and dignity. We should be able to look at digital technology with levelheaded admiration rather than unwarranted awe or suspicion. To make the digital world a world in which we want to live.

Acknowledgments

I wrote *How to Stay Smart in a Smart World* for a general audience to navigate the challenges of a digital world. It builds on two earlier popular books of mine, *Gut Feelings* and *Risk Savvy*. *Gut Feelings* made intuition understandable and scientifically respectable, and *Risk Savvy* helped readers cope with risk and uncertainty. *How to Stay Smart* is a guide to staying in the driver's seat in a world increasingly populated by algorithms. It is not an academic book, but it draws heavily on research, including my own on decision-making under uncertainty at the Max Planck Institute for Human Development in Berlin. I am lucky to have the generous and unique support of the institute and to profit from its splendid intellectual atmosphere. It's research paradise. I also would like to thank David and Claudia Harding for their long-term support of the Harding Center for Risk Literacy, which I direct, now located at the University of Potsdam. For those who wish to learn more about the underlying research, I recommend *Classification in the Wild* (Katsikopoulos, Simsek, Buckman, and Gigerenzer, 2020, MIT Press) and *Simply Rational: Decision Making in the Real World* (Gigerenzer, 2015, Oxford University Press). Further pointers to the scholarly literature are provided in the reference list of this book.

I would like to thank Tom Chan, Eli Finkel, Wolfgang Gaissmaier, Sophie Hartmann, Gisela Henkes, Ralph Hertwig, Konstantinos Katsikopoulos, Gary Klein, Brenden Lake, Shenghua Luan, Sarah McGrew, John Monahan, Jean Czerlinski Ortega, Felix Rebitschek, Raul Rojas, Cynthia Rudin, Natasha D. Schüll, Katharina Schüller, Stephen Shladover, Özgür Şimşek, Amy Slep, Isaac Stanley-Becker, Riccardo Viale, Sam Wineburg, Jason Yosinski, and John Zerilli for their many helpful comments. My special thanks go to Rona Unrau, who edited the entire manuscript and helped me find sources and to improve draft after draft. Rona has been a wonderful support. I am also thankful to Sarah Otterstetter, who designed the figures, and to Rachel Fudge from MIT Press for her keen eye for detail. Last but not least, my thanks go to my wife Lorraine Daston, my daughter Thalia Gigerenzer, and my son-in-law Kyle Chan, who provided intellectual input and emotional support over the years spent working on this book, including the extraordinary year of 2020. The coronavirus put a stop to most of my scheduled talks and travel, but in my case the cloud had a silver lining: I suddenly had more time to work on this book.

Notes

Introduction

1. For instance, throughout her seminal analysis of surveillance capitalism, Shoshana Zuboff appears to assume this ability as generally true, such as when she speaks of "monitoring and shaping of human behavior with unprecedented accuracy" ("Surveillance Capitalism," 17) or writes that "surveillance capitalists know everything *about us*" (*Age of Surveillance Capitalism*, 11).

2. Breakstone et al., "Students' Civic Online Reasoning." See also McGrew et al., "Can Students Evaluate Online Sources?"; Wineburg and McGrew, "Lateral Reading;" and chapter 11.

3. Rogers, "Census in Pictures."

4. National Consortium for the Study of Terrorism and Responses to Terrorism, *Global Terrorism Overview*.

5. Centers for Disease Control and Prevention, "Transportation Safety."

6. Britt, "Drivers on Cell Phones."

7. ERGO, *Risiko-Report*.

8. https://www.youtube.com/watch?v=t7911kgJJZc.

9. For instance, arguing in favor of mass surveillance, the head of the German police trade union wrote that the false positive rate of 0.1 percent is so low that it is "acceptable, barely perceptible." The press release can be accessed here: https://www.bmi.bund.de/SharedDocs/pressemitteilungen/DE/2018/10/gesichtserkennung-suedkreuz.html.

10. The heated debate was fueled by lack of risk literacy. For instance, one commentator on Twitter correctly pointed out that the 0.1 percent false alarm rate together with the low rate of suspects means that the vast majority of alarms are false. To see this, assume 500 suspects hanging out at train stations, of which the system

correctly identifies 400 (80 percent hit rate). Thus, there are 400 correct positives for every 12,000 false positives, which means that about 97 percent of all positive classifications are wrong. In response, a member of the European Parliament defended the system and tweeted: "You should work on your math: a 0.1 percent error rate means 99.9 percent hits when applied optimally. That's optimal." She wrongly thought that a false alarm rate of 0.1 percent means that 99.9 percent of all classifications are correct. When the error was pointed out to her, she insisted that she knew the probability calculus but that it was irrelevant because the real issue at stake was the preventive and educative effect of face-recognition systems. Politicians need risk literacy—and the humility to admit errors as well.

11. Tolentino, "What It Takes to Put Your Phone Away," 71.

12. As the saying goes, "If you do not pay for a product, you are the product." More precisely, the predictions made about you are the product. The coffee house story was inspired by an entertaining talk on corporate surveillance, see: https://idlewords.com/talks/internet_with_a_human_face.htm.

13. Brin and Page, "Anatomy of a Large-Scale Hypertextual Web Search Engine." The story of how the Google founders—helped by a brilliant Canadian mathematician whose loathing of advertising even exceeded that of Page—ended up creating the most powerful ad-based platform is told by Wu, *Attention Merchants*.

14. Various companies, including Mozilla and Google Contributor, are also experimenting with micropayments (Melendez, "Mozilla and Creative Commons"). Netflix is also interested in keeping us watching in order to keep subscription rates high but not to satisfy a host of advertisers. Some platforms offer the option to pay and not be bombarded with advertisements, but few guarantee that those who pay are no longer tracked. For instance, business platforms such as LinkedIn and Xing offer premium memberships that allow access to data of other members (e.g., who visited one's page) but do not protect one's own data.

15. The issue of boosting users' understanding is discussed in the European Commission's report (Lewandowsky et al., *Technology and Democracy*). Digital risk for children has been addressed by private organizations such as the DQ Institute (Chawla, "Need for Digital Intelligence").

16. More precisely, the slots are auctioned and sold to the bidder for whom an algorithm calculates the highest return for the tech company, which is a function of both the bid and the expected number of users who will see or click the ad. Advertisers pay Google for clicks, so clearly Google wants you to click on the ads.

17. https://www.samsung.com/hk_en/info/privacy/smarttv/; Fowler, "You Watch TV."

18. https://www.searchenginewatch.com/2016/04/27/do-50-of-adults-really-not-recognise-ads-in-search-results/.

19. Kawohl and Becker, *Verfügen Deutsche Vorstände.*

20. The Economist, "What to Make of Mark Zuckerberg's Testimony."

21. Advisory Council for Consumer Affairs at the Federal Ministry of Justice and Consumer Protection, *Digital Sovereignty.* The fact that checks on scoring systems are based on reports that the credit agencies themselves have commissioned was revealed by reporters from the Bavarian public broadcaster Bayerischer Rundfunk.

22. See, for example, the 2045 initiative (http://2045.com/ideology/). Some writers dream of being allowed to upload our brains into this vast intelligence, once it arrives, thereby making them immortal.

23. Daniel and Palmer, "Google's Goal."

24. Overstated claims about algorithms without supporting evidence can be found, for instance, in Harari, *Homo Deus.* I provide examples in chapter 11.

25. See the spectrum of opinions in Brockman, *Possible Minds.* Also, Kahneman ("Comment," 609) poses the question whether AI can eventually do whatever people can do: "Will there be anything that is reserved for human beings? Frankly, I don't see any reason to set limits on what AI can do." And: "You should replace humans by algorithms whenever possible" (610).

26. Gigerenzer, *Risk Savvy.*

27. On fear cycles, see Orben, "Sisyphean Cycle."

Chapter 1

1. Anderson et al., "Upsides and Downsides."

2. Grzymek and Puntschuh, *What Europe Knows and Thinks.*

3. Parship spends about 100 million euros per year on TV, website, and poster advertisements, where professional models pose as singles. Theile, "Parship-Chef."

4. Deutsches Institut für Service-Qualität, "Kundenbefragung."

5. https://www.elitesingles.com.

6. Cited in Finkel et al., "Online Dating," 24.

7. https://www.jdate.com/en/jlife/success-stories/shlomit-ryan.

8. Finkel et al., "Online Dating," 3.

9. Epstein, "Truth About Online Dating." eHarmony personnel presented a research paper claiming that married couples who met through eHarmony are happier than those who met by other means. Yet they compared newlywed eHarmony couples (average six months) with couples in the control group that were married on average

for more than two years. The study neither passed the scientific review process nor was published, and one can guess why.

10. When, for instance, I contacted ElitePartner requesting an interview for this book, I was asked to send in the questions beforehand. I submitted ten questions. The response was that they were not willing to answer half of these, such as how many new paying members they have per year and how many fall in love per year. I am still waiting for their answers to the other half. There is reason to doubt that the figures released by dating agencies are trustworthy; the hyperbolic claims of the matchmaking industry have been analyzed by Zoe Strimpel, "Matchmaking Industry."

11. Cacioppo et al., "Marital Satisfaction."

12. Danielsbacka et al., "Meeting Online"; Potarca, "Demography of Swiping Right"; Paul, "Is Online Better than Offline."

13. Potarca, "Demography of Swiping Right"; Thomas, "Online Exogamy"; Brown, "Couples Who Meet Online."

14. Tinder also uses an algorithm, albeit one that is not based on profiles. Tinder's algorithm gives each user a secret desirability score based on how many swipes they get and from whom.

15. https://www.youtube.com/watch?v=m9PiPlRuy6E.

16. To make it simple, I use the arithmetic mean here. OKCupid's algorithm calculates the *geometric mean*, which multiplies the values and takes the square root.

17. Bruch and Newman, "Aspirational Pursuit of Males."

18. Rudder, *Dataclysm*.

19. Buss, *Evolutionary Psychology*.

20. Finkel et al., "Online Dating," 30.

21. Joel et al., "Is Romantic Desire Predictable?"

22. Todd et al., "Different Cognitive Processes." See also Finkel et al., "Online Dating."

23. Montoya et al., "Is Actual Similarity Necessary."

24. Dyrenforth et al., "Predicting Relationship and Life Satisfaction."

25. Finkel et al., "Online Dating," 44.

26. Sales, *American Girls*.

27. Epstein, "Truth About Online Dating."

28. Hancock et al., "Truth About Lying."

29. Hitsch et al., "Matching and Sorting in Online Dating."

30. Epstein, "Truth About Online Dating."

31. Epstein, "Truth About Online Dating."

32. Anderson, "Ugly Truth About Dating."

33. Epstein, "Truth About Online Dating."

34. Lea et al., *Psychology of Scams*, 42.

35. Facebook prohibits misleading advertising, but scammers use a technique called "cloaking" that shows Facebook's robots harmless content, unlike the message the victim reads (Kayser-Bril, "Facebook Enables Automated Scams").

36. Office of Fair Trading, United Kingdom, *Impact of Mass Marketed Scams*.

37. Whitty and Buchanan, "Online Dating Romance Scam."

38. Suarez-Tangil et al., "Automatically Dismantling Online Data Fraud."

39. Whitty and Buchanan, "Online Dating Romance Scam."

40. Tsvetkova et al., "Even Good Bots Fight."

41. Brown, "24 Victims of the Ashley Madison Hack."

42. Federal Trade Commission, "FTC Sues Owner of Online Dating Service."

43. Bostrom, *Superintelligence*, 211.

44. Youyou et al., "Computer-Based Personality Judgments."

Chapter 2

1. Attributed to Richard Feynman while speaking at a Caltech graduation ceremony (source of the quote unclear).

2. Cited by Boden, *Mind as Machine*, 840. A puzzling tension exists between Simon's work on decision-making and bounded rationality, which focuses on uncertainty, and his work on AI, which focuses on well-defined problems such as chess. In his work on bounded rationality, Simon distinguishes between uncertainty and well-defined problems, while in his writings on AI this distinction disappears (Gigerenzer, "What Is Bounded Rationality?").

3. Dreyfus, *What Computers Can't Do*, 33.

4. See Katsikopoulos et al., *Classification in the Wild*, where it is called the "unstable world principle," which is the same principle.

5. Kay and King, *Radical Uncertainty*; Taleb, *Black Swan*.

6. Makoff, "Computer Wins on 'Jeopardy!.'"

7. Ferrucci et al., "Building Watson."

8. Lee, *AI Superpowers*.

9. Russell, *Human Compatible*, 47–48.

10. Kumar et al., "Attribute and Simile Classifiers." In this experiment, 50 percent of the presented pairs showed the same person, and the other 50 percent showed different people.

11. For this and the following example, see Katsikopoulos et al., *Classification in the Wild*, chap. 4.

12. Chiusi, "Life in the Automated Society," 195.

13. Simon and Newell, "Heuristic Problem Solving."

14. Krauthammer, "Be Afraid."

15. See Katsikopoulos et al., *Classification in the Wild*.

16. Aikman et al., "Taking Uncertainty Seriously."

17. Gigerenzer et al., *Heuristics;* Katsikopoulos et al., *Classification in the Wild*.

18. Wachter, *Digital Doctor*.

19. Kellermann and Jones, "What It Will Take."

20. Kellermann and Jones, "What It Will Take."

21. Wachter, *Digital Doctor*.

22. Schulte and Fry, "Death by 1,000 Clicks."

23. See Carr, *Glass Cage*.

24. Young et al., "*A Time-Motion Study*."

25. Wachter, *Digital Doctor*.

26. Schulte and Fry, "Death by 1,000 Clicks."

27. AFP, "'Shocking' Hack of Psychotherapy Records."

28. Wachter, *Digital Doctor*, 71.

29. Gigerenzer and Muir Gray, *Better Doctors*.

30. Gottman and Gottman, "Natural Principles of Love," 10; https://www.gottman.com/about/research/; Gottman et al., "Predicting Marital Happiness."

31. See Heyman and Slep, "Hazards of Predicting Divorce."

32. https://www.gottman.com. See also Barrowman, "Correlation, Causation, and Confusion."

33. Roberts and Pashler, "How Pervasive Is a Good Fit?"

34. Heyman and Slep, "Hazards of Predicting Divorce."

35. Heyman and Slep, "Hazards of Predicting Divorce."

36. Heyman and Slep, "Hazards of Predicting Divorce." The difference between 65 percent and 21 percent is called *overfitting*.

37. For the field of behavioral economics, see for example, Berg and Gigerenzer, "As-If Behavioral Economics."

38. Fitting was also the rule in statistics, such as in the *Journal of the American Statistical Association*; see Breiman, "Statistical Modeling."

39. Until the 1990s, one would have had to search hard to find any predictions in cognitive modeling in general. See Roberts and Pashler, "How Pervasive Is a Good Fit?" and Brandstätter et al., "Risky Choice with Heuristics."

40. Bailey et al., "Pseudo-Mathematics and Financial Charlatanism."

41. https://quoteinvestigator.com/tag/niels-bohr/.

42. Russell and Norvig, *Artificial Intelligence*.

43. Strickland, "How IBM Watson Overpromised and Underdelivered."

44. Ross and Swetlitz, "IBM Watson Health Hampered."

45. Topol, *Deep Medicine*.

46. Best, "IBM Watson."

47. Schwertfeger, "Künstliche Intelligenz."

48. Brown, "Why Everyone Is Hating on IBM Watson."

Chapter 3

1. Gigerenzer and Goldstein, "Mind as Computer"; Gigerenzer and Murray, *Cognition as Intuitive Statistics*.

2. Daston, "Enlightenment Calculations"; Gigerenzer and Goldstein, "Mind as Computer."

3. Babbage was similarly influenced by the division of labor introduced in the English textile industry that used punch cards to program weaving machines. This so

impressed Babbage and many of his contemporaries that "factory tourism" came into vogue. See Daston, "Enlightenment Calculations," and Gigerenzer, "Digital Computer."

4. There are many versions of this story. This condensed version is based on one of the first accounts that mentions the numbers 1 to 100 by Franz Mathé in 1906. See Brian Hayes, "Gauss's Day of Reckoning."

5. Cited in Wood, "He Made the Mini." This quote has also been attributed to another Sir Alec, namely Alec Guinness. See also the comments by former US President Barack Obama (Martosko, "I'm Bad at Math, Says Obama").

6. Gleick, *Genius*.

7. Daston, "Calculation and the Division of Labor."

8. "Women in Computer Science."

9. von Neumann, *Computer and the Brain*; Turing, "Computing Machinery." On the differences between von Neumann's, Turing's, and Simon's view of the relation between the computer and the brain or mind, see Gigerenzer and Goldstein, "Mind as Computer."

10. Newell et al., "Elements of a Theory."; Newell and Simon, *Human Problem Solving*.

11. Simon uses clear words: "the hypothesis is that a physical symbol system I have just described has the necessary and sufficient means for general intelligence" (*Sciences of the Artificial*, 28).

12. Simon, *Models of My Life*.

13. Cohen, "Howard Aiken on the Number of Computers."

14. Gigerenzer and Goldstein, "Mind as Computer."

15. Gigerenzer, "From Tools to Theories"; Gigerenzer and Murray, *Cognition as Intuitive Statistics*.

Chapter 4

1. Uber sold its autonomous car division (Uber ATG) in December 2020.

2. National Transportation Safety Board, *Preliminary Report Highway*; Stern, "Self-Driving Uber Crash."

3. Stilgoe, "Who Killed Elaine Herzberg?"

4. Efrati, "How an Uber Whistleblower Tried to Stop."

5. Stilgoe, "Who Killed Elaine Herzberg?"

6. In 2016, for instance, Nissan CEO Carlos Ghon proclaimed regarding his cooperation with Microsoft, "In 2020, you're going to have what we call a totally autonomous-driven car," before adding: "This is different from the driverless car . . . 2020 for the autonomous car in urban conditions, probably 2025 for the driverless car" (Dillet, "Renault-Nissan CEO Carlos Ghosn." See also Elias, "Alphabet Exec.")

7. Dickmanns and Zapp, "Autonomous High Speed Road Vehicle Guidance."

8. Shladover, "Truth About 'Self-Driving' Cars."

9. Shladover, "Truth About 'Self-Driving' Cars."

10. Daston, *Rules*.

11. See Geirhos et al., "Shortcut Learning."

12. Simonite, "When It Comes to Gorillas, Google Photos Remains Blind."

13. Szegedy et al., "Intriguing Properties of Neural Networks."

14. Here are some of statisticians' complaints about the new jargon in AI. What is now called the *input layer* is known as *independent variables* in statistics, and the *output layer* is a new term for *dependent variables*. What is called a *neural network* is in essence the same as a nonlinear regression or a discriminant analysis, for numerical estimation and classification, respectively. Moreover, algorithms that can learn were developed long ago. One of these is *regression*, which learns weights from data, as in supervised learning. In statistics, unsupervised learning also has a long tradition, with a toolbox of methods such as cluster analysis and factor analysis. Despite this similarity between classical statistics and AI, one key difference is that statistical applications typically fit their models to a given sample of data, whereas neural networks get the data *sequentially* and update their algorithm after each feedback. See Sarle, "Neural Networks."

15. Kosko, "Thinking Machines."

16. Szegedy et al., "Intriguing Properties of Neural Networks."

17. These tiny systematic perturbations that fool deep neural networks can be found with the help of evolutionary algorithms. The technique is also called *hard-negative mining*, that is, identifying "adversarial examples" to which the network falsely attaches low probabilities of being the correct number or object. The same perturbation can cause misclassifications not only for the examples shown here but also for other training sets and with other networks. To shield networks, one strategy has tried to include adversarial examples in the training set, but even that method remains vulnerable to simple attacks (Tramer et al., "Ensemble Adversarial Training").

18. Wang et al., "Towards a Robust Deep Neural Network."

19. The images that fool deep neural networks were generated with evolutionary algorithms or gradient ascent methods (Nguyen et al., "Deep Neural Networks Are Easily Fooled").

20. Su et al., "One Pixel Attack." On the logic of adversarial networks, see Rocca, "Understanding GANs."

21. Brooks, "Mistaking Performance."

22. Titz, "Deep Neural Networks."

23. Ranjan et al., "Attacking Optical Flow."

24. Dingus et al., "Driver Crash Risk Factors."

25. Webb, *Big Nine*, 183. To deal with these problems, engineers are working on multisensory systems, which use a combination of vision, laser (lidar), and location measures. For instance, the location is linked to a map with information about speed limits, so that a 65-mile-per-hour speed limit sign "detected" by the vision system could be disregarded.

26. Gleave, "Physically Realistic Attacks."

27. Luetge, "German Ethics Code."

28. Awad et al., "Moral Machine Experiment."

29. Fleischhut et al., "Moral Hindsight."

30. Steven E. Shladover, personal communication, December 2020.

31. https://www.sparkassen-direkt.de/auto-mobilitaet/telematik/. The example is taken from the first German telematics insurer, Sparkassen.direkt. Rapid acceleration is defined as exceeding 0.235 g = 2.3m/sec^2, and harsh braking as exceeding 0.286 g = 2.8m/sec^2. The company eventually discontinued offering telematics insurance because the cost of the black box was too high. Nevertheless, it became a financial success but for other reasons: its publicity attracted new customers. Due to new regulations in the European Union, all new cars must have an e-call system, that is, a black box that notifies the police in the case of a severe car accident. This regulation makes telematics insurance more profitable.

32. ERGO, *Risiko-Report*. The study used a representative sample of 3,200 people living in Germany.

33. Medvin, *Your Vehicle Black Box*.

34. Doll, "Moderne Autos."

35. Fowler, "What Does Your Car Know About You?"

36. Maack, "Diese Stadt."

37. Bliss, "How Utrecht Became a Paradise."

38. National Geographic, "Greendex 2014."

39. Branwen, "Neural Net Tank Urban Legend."

40. See Geirhos et al., "Shortcut Learning."

41. Szabo, "Artificial Intelligence."

42. Zech et al., "Variable Generalization Performance."

43. See, for instance, Youyou et al., "Computer-Based Personality Judgments."

44. Koh, "WILDS"; Geirhos et al., "Shortcut Learning."

Chapter 5

1. Church, "Rights of Machines." The 85 billion nerve cells in the brain consume about 20 percent of the body's energy; thus, the average power consumption of an adult is 100 watts.

2. Tomasello, *Becoming Human*.

3. Gigerenzer, *Gut Feelings*.

4. Kruglanski and Gigerenzer, "Intuitive and Deliberate Judgments."

5. Judea Pearl is possibly the most prominent researcher who focuses on causal reasoning in artificial intelligence, yet we are nowhere near to adding this ability to algorithms.

6. Poibeau, *Machine Translation*.

7. For illustrations of translations from Chinese, French, and German into English, see Hofstadter, "Shallowness of Google Translate."

8. https://www.deepl.com/en/translator#de/en/Papstschuss. The information is fluid on DeepL, meaning that these translations may change with time.

9. Poibeau, *Machine Translation*.

10. Poibeau, *Machine Translation*, 71.

11. Kayser-Bril, "Female Historians and Male Nurses."

12. Niven and Kao, "Probing Neural Network Comprehension." Similarly, Gururangan et al. ("Annotation Artifacts") conclude that because of neural networks' ability to discover spurious but efficient cues in the data, the success of natural language models to date has been overestimated.

13. Kurzweil, *How to Create a Mind*, 7.

14. Quinn et al., "Perceptual Categorization."

15. Sinha et al., "Face Recognition by Humans."

16. Brunswik, *Perception and Representative Design*.

17. Sinha and Poggio, "I Think I Know that Face."

18. Asch, "Effects of Group Pressure."

19. Szegedy et al., "Intriguing Properties."

20. Burell, "How the Machine 'Thinks.'"

21. Lake et al., "Ingredients of Intelligence."

22. Nguyen et al., "Deep Neural Networks Are Easily Fooled."

23. Lake et al., "Ingredients of Intelligence."

24. Hsu, "Transasia Pilot Shut Down Wrong Engine."

25. Luria, *Mind of a Mnemonist*.

Chapter 6

1. Daston, "Immortal Archive."

2. Daston, "Immortal Archive," 161–162.

3. Google engineers trained the algorithms on data from the years 2003–2007 and tested them on data from 2007–2008. Ginsberg et al., "Detecting Influenza Epidemics."

4. Olson et al., "Reassessing Google Flu Trends Data."

5. Lazer et al., "Parable of Google Flu."

6. Copeland et al., "Google Disease Trends."

7. Nevertheless, bestselling authors continue to suggest to the general public that Google Flu Trends, like IBM's Watson for Oncology, is a huge success story for big data. I recall a talk held in Munich in May 2014 by Victor Meyer-Schönberger, first author of the 2013 bestseller *Big Data*, in which he still presented Google Flu Trends as the showcase of big data even though by that time, Google Flu Trends had long lost its nose for where the flu was going. The hundreds of business executives in the audience were visibly impressed. Two years later, Harari (*Homo Deus*, 390–391) still talked about Google's "magic" and presented the story in a similar way to impress the general reader.

8. Anderson, "End of Theory."

9. Brown, *Philosophy of the Human Mind*.

10. Katsikopoulos et al., *Classification in the Wild* and "Transparent Modeling."

11. Katsikopoulos et al., *Classification in the Wild* and "Transparent Modeling". Our analysis was inspired by the appendix in Lazer et al. ("Parable of Google Flu"), who reported various more or less complex rules that predict better than Google Flu Trends. For those interested in the statistical details, note that these rules included a simple regression with a two-week lag as the only predictor that had the same error as the recency heuristic. The regression model differs from the recency heuristic in that one still needs to compute and update the regression coefficients every time one makes a prediction. This is no improvement over the recency heuristic. Lazer reported that a hybrid model that combines Google Flu Trends with lag data performs best. Yet this hybrid model needs all elaborate calculations performed by Google Flu Trends in the first place.

12. See Artinger et al., "Recency"; Green and Armstrong, "Simple Versus Complex Forecasting"; and Katsikopoulos et al., "Transparent Modeling."

13. Dosi et al., "Rational Heuristics."

14. Porter, *Karl Pearson.*

15. Anderson, "End of Theory."

16. Ginsberg et al., "Detecting Influenza Epidemics."

17. Vigen, *Spurious Correlations.*

18. Aschwanden, "You Can't Trust what You Read."

19. Advisory Council for Consumer Affairs at the Federal Ministry of Justice and Consumer Protection, *Digital Sovereignty*, 62. In the financial world, being fooled by randomness is a frequent event; see Taleb, *Fooled by Randomness.*

20. Mullard, "Reliability of 'New Drug Target'"; Begley, "In Cancer Science."

21. Chalmers and Glasziou, "Avoidable Waste."

22. Leetaru, "Data Brokers so Powerful."

23. Bergin, "How a Data Mining Giant Got Me Wrong."

24. Anderson, "What Happens to Big Data."

25. Neumann et al., "Frontiers."

26. Salganik et al., "Measuring the Predictability of Life Outcomes."

27. Clayton et al., *Machine Learning.*

28. Chiusi, "Life in the Automated Society." For instance, Danish municipalities had been experimenting with an algorithm for identifying children in vulnerable families until the data protection authorities intervened.

29. Instituto Superiore di Sanità, *Characteristics of COVID-19 Patients*. The report is based on available data from November 18, 2020.

30. Gigerenzer, *Risk Savvy*.

Chapter 7

1. *Loomis v. Wisconsin*, 881 N.W.2d 749 (Wis. 2016), cert. denied, 137 S.Ct. 2290 (2017).

2. Liptak, "Sent to Prison."

3. See Gigerenzer, *Calculated Risks*, 186.

4. Grisso and Tomkins, "Communicating Violence Risk Assessments," 928.

5. The study analyzed more than 1,000 rulings of eight Israeli judges (Danziger et al., "Extraneous Factors"). For a critique of its conclusions, see Weinshall-Margel and Shapard ("Overlooked Factors") and Glöckner ("Irrational Hungry Judge Effect").

6. DeMichele et al., "Public Safety Assessment."

7. Epstein, *Simple Rules*.

8. Dressel and Farid, "Accuracy, Fairness, and Limits."

9. Dressel and Fari, "Accuracy, Fairness, and Limits."

10. Matacic, "Are Algorithms Good Guides?"

11. This is the average of the value for individuals (63%) and the value of the majority vote in groups of twenty (67%).

12. Geraghty and Woodhams, "Predictive Validity of Risk Assessment Tools." The Dressel and Farid study was replicated by Lin et al. ("Limits of Human Prediction") with similar results, except when the base rates of recidivism were low (only 11%) *and* the human participants did not get feedback and thus could not learn about the low base rates. With low base rates of recidivism, COMPAS fared no better than the most simple algorithm of always predicting no recidivism.

13. See DeMichele et al., "Public Safety Assessment," and Rudin et al., "Age of Secrecy and Unfairness."

14. Angelino et al., "Learning Certifiably Optimal Rule Lists."

15. Rudin and Radin, "Why Are We Using Black Box Models." If you are interested in reading more on how machine-learning tools extract the most important features, see Katsikopoulos et al., *Classification in the Wild*. The definition of transparency in this section is also taken from this work.

16. https://advancingpretrial.org/psa/factors/#fta.

17. My colleagues and I have worked on making decision lists more transparent by curtailing their size, which, in machine learning, is typically unrestricted (Katsikopoulos et al., *Classification in the Wild*).

18. https://advancingpretrial.org/psa/research/.

19. See Stevenson, "Assessing Risk Assessment," for the problems and traps of measuring the performance of algorithmic risk assessment tools.

20. Grossman et al., "50 Best Inventions."

21. Peteranderl, *Under Fire*.

22. Gorner and Sweeney, "For Years Chicago Police Rated the Risk."

23. Peteranderl, *Under Fire*.

24. Thomas, "Why Oakland Police Turned Down Predictive Policing."

25. Hauber et al., *Prädiktionspotenzial*.

26. Snook et al., "Complexity and Accuracy of Geographic Profiling Strategies."

27. See Bennell et al., "Clinical Versus Actuarial Geographic Profiling," and Paulsen, "Human Versus Machine."

28. Gigerenzer et al., *Heuristics*.

29. Sergeant and Himonides, "Orchestrated Sex."

30. Dastin, "Amazon Scraps Secret AI Recruiting Tool." See O'Neill, *Weapons of Mass Destruction*, for how big data increases inequality.

31. Buolamwini and Gebru, "Gender Shades."

32. Raji and Buolamwini, "Actionable Auditing."

33. Zhao et al., "Men Also Like Shopping." This article also developed measures that succeeded in reducing the increase in gender bias by almost half but couldn't fully eliminate it.

34. Among the two-thirds of pictures classified as "woman," two-thirds are expected to be correct (because two-thirds of all the pictures are women). Similarly, among the one-third of the pictures classified as "man," one-third are expected to be correct (because only one-third of all the pictures are men), which results in $2/3 \times 2/3 + 1/3 \times 1/3 = 5/9$ correct answers, about 56 percent.

35. Chin, "AI is the Future." Google says that over 20 percent of technical roles are filled by women, but that is not the figure for machine learning.

36. Hao, "Paper that Forced Timnit Gebru out of Google."

37. https://googlewalkout.medium.com/standing-with-dr-timnit-gebru-isupport timnit-believeblackwomen-6dadc300d382.

38. Gigerenzer et al., "Helping Doctors and Patients."

39. https://www.hardingcenter.de/en.

40. Benoliel and Becher, "Duty to Read the Unreadable."

41. McDonald and Cranor, "Cost of Reading Privacy Policies."

42. Zuboff, *Age of Surveillance Capitalism*, 166–168.

43. Advisory Council for Consumer Affairs at the Federal Ministry of Justice and Consumer Protection, *Digital Sovereignty*.

44. Rudin and Radin, "Why Are We Using Black Box Models."

45. https://www.fico.com/en/newsroom/fico-announces-winners-inaugural-xml -challenge.

46. Rudin, "Stop Explaining Black Box Machine Learning Models." For the medical domain, see Holzinger et al., "What Do We Need to Build Explainable AI."

47. Katsikopoulos et al., *Classification in the Wild*.

48. Turek, *Explainable Artificial Intelligence*, 7–10. See also Gunning and Aha, "DARPA's Explainable Artificial Intelligence Program."

49. See for instance, Gigerenzer et al., *Heuristics*; Green and Armstrong, "Simple Versus Complex Forecasting"; Jung et al., "Simple Rules"; and Katsikopoulos et al., *Classification in the Wild* and "Transparent Modeling."

50. Artinger et al., "Recency"; Wübben and von Wangenheim, "Instant Customer Base Analysis: Managerial Heuristics Often 'Get It Right.'"

51. In machine learning, these rules are known as 1-rules, learned by the 1R program; see Holte, "Very Simple Classification Rules."

52. Rivest, "Learning Decision Lists."

53. The fact that tallying can be just as or even more accurate than black box algorithms in situations of uncertainty has been known in psychology since Dawes and Corrigan ("Linear Models"), Einhorn and Hogarth ("Unit Weighting Schemes"), and Czerlinski et al. ("How Good Are Simple Heuristics?"). A comparison between tallying and machine-learning algorithms can be found in Katsikopoulos et al., *Classification in the Wild*.

54. Markman, "Big Data."

55. Silver, "Why Our Model Is More Bullish."

56. Lichtman, *Predicting the Next President*. The following presentation is adapted from Katsikopoulos et al., *Classification in the Wild*.

57. Stevenson, "Trump Is Headed for a Win."

58. In the 2020 US presidential election, the Keys predicted correctly that Biden would win. The polls again underestimated Trump, but most got the final result right.

59. Spinelli and Crovella, "How YouTube Leads Privacy-Seeking Users."

60. Pasquale, *Black Box Society*.

61. See Gigerenzer, *Calculated Risks*, 91.

62. Advisory Council for Consumer Affairs at the Federal Ministry of Justice and Consumer Protection, *Consumer-Friendly Scoring*. The members of the council were divided about whether scoring services should report all features or only the most relevant to the public.

63. See Articles 13 and 22 of the GDPR (European Union).

64. https://docs.google.com/forms/d/e/1FAIpQLSfdmQGrgdCBCexTrpne7KXUzpbi I9LeEtd0Am-qRFimpwuv1A/viewform.

Chapter 8

1. Nasiopoulos et al. ("Wearable Computing") report that wearing smart glasses increases socially accepted behavior, but only as long as people do not forget about the eye trackers. On the positive effect of security cameras on prosocial behavior, see Rompay et al., "Eye of the Camera."

2. Cho, "Social Credit System." On Venezuela, see Berwick, "How ZTE Helps Venezuela."

3. Kostka and Antoine, "Fostering Model Citizenship."

4. Kostka, "China's Social Credit System."

5. Advisory Council for Consumer Affairs at the Federal Ministry of Justice and Consumer Protection, *Consumer-Friendly Scoring*.

6. ERGO, *Risiko-Report*.

7. Rötzer, "Jeder Sechste Deutsche."

8. Christl, *Corporate Surveillance*.

9. Leetaru, "Data Brokers So Powerful."

10. Barnes, "Privacy Paradox."

11. Vodafone Institut für Gesellschaft und Kommunikation, "Big Data."

12. Norton LifeLock, *Cyber Safety Insights Report*.

13. ERGO, *Risiko-Report.*

14. Freed et al., "Stalker's Paradise."

15. In *The Circle*, Dave Eggers's satire of Silicon Valley, a cult-like internet company—a mixture of Google and Facebook—reigns the world. At company rallies, employees chant "sharing is caring" and "privacy is theft."

16. Farnham, "Hot or Not's Co-Founders."

17. Johnson, "Privacy No Longer a Social Norm."

18. "What to Make of Mark Zuckerberg's Testimony."

19. Zuboff, "You Are Now Remotely Controlled."

20. Brin and Page, "Anatomy of a Large-Scale Hypertextual Web Search," 3832.

21. Zuboff, *Age of Surveillance Capitalism*, 71–75.

22. Zuckerman, "Internet's Original Sin."

23. Zuboff, *Age of Surveillance Capitalism.*

24. Zuboff *Age of Surveillance Capitalism*, 87.

25. Zuboff, *Age of Surveillance Capitalism*, 118–120.

26. Tech Transparency Project, "Google's Revolving Door (US)," https://www.tech transparencyproject.org/articles/googles-revolving-door-us.

27. Baltrusaitis, "Top 10 Countries and Cities."

28. Elliott and Meyer, "Claim on 'Attacks Thwarted' by NSA."

29. Cahall et al., "Do NSA's Bull Surveillance Programs."

30. For more details, see Zuboff, *Age of Surveillance Capitalism.*

31. Watson, "Humans Have Shorter Attention Span."

32. Smith et al., "2010 Survey on Cell Phone Use."

33. Twenge, *iGen*. I should note that the questions of whether these associations are causal and how large the effect is are a matter of intense discussion in psychology. See also Turkle, *Reclaiming Conversation.*

34. Lanier and Weyl, "Blueprint for a Better Digital Society"; Lanier, "Jaron Lanier Fixes the Internet."

35. Here I have updated a calculation made by Felix Stalder, "Paying Users."

36. Noyes, *Top 20 Valuable Facebook Statistics.*

37. Facebook, "Fourth Quarter and Full Year 2019 Results."

38. For other estimates, see Munro, "Should Tech Firms Pay People." Some higher estimates are based on dividing Facebook's revenue by the number of users, which ignores that fact that revenue is not income, and also that Facebook needs to retain some of the income for profit.

39. Melendez, "Mozilla and Creative Commons"; Pogue, "How to End Online Ads."

40. Eames, *Barbie Doll Fashion.* Mattel has sold more than a billion Barbies.

41. Dittmar et al., "Does Barbie Make Girls Want to Be Thin?"

42. K. Bondy, cited in Gigerenzer, *Calculated Risks*, 23, 260.

43. Praschl, "Eine Barbie mit WLAN."

44. Digital Courage, "'Stasi-Barbie.'" Other companies followed and produced their own smart dolls, such as My Friend Cayla, a toy that was classified as an illegal transmitter and forbidden in Germany. Illegal transmitters are well-known from Cold War movies where spies record private conversations using pens and cigarette lighters with hidden microphones built in.

45. Hu, "How One Light Bulb Could Allow Hackers."

46. Tzezana, "Scenarios for Crime and Terrorist Attacks."

47. University of Michigan, "Hacking into Home." See also Davis et al., "Vulnerability Studies."

48. Elsberg, *Blackout.*

49. https://www.samsung.com/hk_en/info/privacy/smarttv/. In 2015, Samsung TVs were found to be listening to everything they could hear. Since then, the company maintains it has changed the system to record only what you say when spoken to directly. The pattern is familiar: companies continually test how far they can go, that is, to what extent the public's privacy expectations have deteriorated, and only react after protest. In 2017, the Federal Trade Commission rebuked the TV industry for being deceptive and unfair and told the industry to be upfront and give the customers a chance to opt in. The industry countered with even more small print that few will take the time to read; the majority will simply click "agree." Fowler, "You Watch Your TV"; Nguyen, "If You Have a Smart TV." The survey mentioned was conducted with the insurer ERGO in Germany in 2019.

50. Wang, "No, the Government Is Not Spying on You Through Your Microwave."

51. Weiser, "Computer for the Twenty-First Century."

52. The best account of nudging and its underlying philosophy can be found in Rebonato, *Taking Liberties.*

53. Epstein and Robertson, "Search Engine Manipulation Effect."

54. Epstein and Robertson, "Search Engine Manipulation Effect"; Zweig, "Watching the Watchers."

55. Ten percent undecided voters is the average of the 2016 and 2012 US presidential elections. Silver, "Where Are the Undecided Voters?"

56. Bond et al., "A 61-Million-Person Experiment."

57. See, for instance, Zuboff, *Age of Surveillance Capitalism* and "Surveillance Capitalism and the Challenge of Collective Action."

58. Coppock et al., "Small Effects of Political Advertising." See also Howard, *Lie Machines*.

59. Kramer et al., "Experimental Evidence of Massive-Scale Emotional Contagion." As an aside, the editor of the journal in which the Facebook article was published explicitly expressed ethical concerns.

60. Booth, "Facebook Reveals News Feed Experiment"; McNeal, "Facebook Manipulated User News Feeds."

61. Snowden, *Permanent Record*.

62. Zuckerberg, "We Need to Rely on and Build More AI Tools." See also Morozov, *To Save Everything*.

63. Shahin and Zheng, "Big Data and the Illusion of Choice."

64. Kostka, "China's Social Credit System."

65. Zuckerman, "Internet's Original Sin."

66. Greenwald, *No Place to Hide*.

67. Snowden, *Permanent Record*.

68. Snowden, *Permanent Record*.

69. Lotan, "Israel, Gaza, War, & Data."

70. Schmidt, "Google CEO Eric Schmidt on Privacy."

71. Kostka and Zhang, "Tightening the Grip." Shenzhen, a city near Hongkong with over twelve million inhabitants, was the first large city in the world to have its taxis and buses run entirely on electricity.

72. For instance, surveillance cameras hang from rooftops and poles across Ecuador, from the Amazonian jungle to the Galápagos islands. Ecuador has implemented a Chinese system called ECU-911, and other countries in Asia, Europe, and Africa appear to be following suit.

73. The term "Great Firewall" was coined by *Wired* magazine in 1997. See Griffiths, *Great Firewall*.

74. Griffiths, *Great Firewall*, 64.

75. Griffiths, *Great Firewall*, 118. The censored search engine Google.cn went online in January 2006.

76. Isaac, "Facebook Said to Create Censorship Tool."

Chapter 9

1. Skinner distinguished two kinds of intermittent (variable) reinforcement and two kinds of fixed reinforcement, resulting in four schedules of reinforcement: (1) in a *fixed ratio schedule*, a behavior is reinforced after it happened exactly *x* times; (2) in a *variable ratio schedule*, a behavior is reinforced after it happened a variable number of times; (3) in a *fixed interval schedule*, a behavior is reinforced if it occurred after a fixed time; (4) in a *variable interval schedule*, a behavior is reinforced if it occurred after a variable interval of time. The two variable reinforcement schedules create a steady rate of behavior and the variable rate schedule the highest response rate. The fixed interval schedule leads to the lowest response rate, and after a reinforcement, the behavior stops and is only gradually taken up again.

2. Halpern, "Creepy New Wave."

3. Lindström et al., "Computational Reinforcement Learning."

4. See Lindström et al., "Computational Reinforcement Learning."

5. Solon, "Ex-Facebook President Sean Parker."

6. https://www.facebook.com/notes/facebook/facebook-gets-a-facelift/2207967130.

7. Zuboff, *Age of Surveillance Capitalism*, 458.

8. Luckerson, "Here's How Facebook's News Feed."

9. Haynes, "Dopamine, Smartphones, and You."

10. Solsman, "YouTube's AI."

11. Spinelli and Crovella, "How YouTube Leads Privacy-Seeking Users." These authors also show that YouTube's recommender system more likely directs privacy-seeking users (e.g., those who disable cookie placements or use Tor) toward unreliable sources than those logged into a Google account. YouTube's policy change in 2019 has somewhat reduced this "leading away" effect, but it remains in force.

12. Note, however, that there is no official score; all is self-reported (different sites list entirely different winners: https://www.distractify.com/p/longest-snapchat-streak). With so many daily active users and so many snaps sent every single day, keeping track of the longest streak becomes a troublesome business. This could be easily

solved by an official scoreboard within the app, but until that happens users all over the world have to keep track themselves.

13. 99 Content, "Snapchat Statistics."

14. Tanz, "Curse of Cow Clicker."

15. Fletcher, "50 Worst Inventions."

16. Eyal, *Hooked*, 175.

17. Zendle and Cairns, "Video Game Loot Boxes."

18. NHS, "Country's Top Mental Health Nurse."

19. Gibbs, "Apple's Tim Cook."

20. Ward et al., "Brain Drain."

21. Newport, "Most U. S. Smartphone Owners."

22. Schüll, *Addiction by Design*, 215.

23. Schüll, *Addiction by Design*, 109.

24. AddictionResource.net, "Average Cost of Illegal Street Drugs."

25. Schüll *Addiction by Design*, 58. Besides video poker, multiline video slots also exemplify the escape-gambling, time-on-device type of play.

26. Schüll, *Addiction by Design*, 33.

27. Schüll, *Addiction by Design*, 2.

Chapter 10

1. Smith, "We Worked Together on the Internet."

2. Centers for Disease Control and Prevention, "Transportation Safety."

3. Kleinhubbert, "Abgelenkt."

4. Gendron, "Croisade Contre les Textos."

5. http://www.txtresponsibly.org/share-your-stories/.

6. Kleinhubbert, "Abgelenkt."

7. Tombu and Jolicoeur, "Virtually No Evidence."

8. Ophir et al., "Cognitive Control."

9. Watson and Strayer, "Supertaskers." The citation is from the abstract.

10. Gliklich et al., "Patterns of Texting and Driving."

11. "Flying to Distraction."

12. Kim, "Are Smartphones Causing More Kid Injuries?"

13. Connolly, "Child Drowning in Germany."

14. Allen, "Sean Parker Unloads on Facebook."

15. For an overview of studies reported here, see McDaniel, "Parent Distraction with Phones."

16. Reed et al., "Learning on Hold."

17. Zimmerman et al., "Associations Between Media Viewing and Language Development."

18. Clifford et al., "Effortful Control." Kozyreva et al. ("Citizens Versus the Internet") review research on digital self-control.

19. BEA, Final Report AF 447, p. 185, as cited in Hartmann, "Risikobewältigung."

20. Federal Aviation Administration, "Safety Alert." See also Carr, *Glass Cage*, 1–2.

21. Brumfiel, "U.S. Navy Brings Back Navigation."

22. Brainbridge, "Ironies of Automation."

23. Ruginski et al., "GPS Use."

24. Baraniuk, "GPS Error."

25. Milner, "Death by GPS"; CNS/ATM, *Resource Guide*.

26. Woollett and Maguire, "Acquiring 'the Knowledge' of London's Layout."

Chapter 11

1. Locke, *Essay Concerning Human Understanding*, 656–657. Italics in original.

2. Topsell, *History of Four-Footed Beasts*.

3. Simanek, "Arthur Conan Doyle."

4. Newton, "Partisan War over Fact-Checking."

5. Götz et al., "Warum Kann Ich Nicht."

6. Lavin, "QAnon, Blood Libel."

7. Islam et al., "COVID-19-Related Infodemic."

8. O'Toole, *Heroic Failure*.

9. O'Toole, *Heroic Failure*.

10. Carroll, *Hunting of the Snark*. On repetition and persuasion in general, see Armstrong, *Persuasive Advertising*.

11. Gigerenzer, "Frequency-Validity Relationship."

12. Mønsted et al., "Evidence of Complex Contagion."

13. See Gigerenzer et al., "Eine Stunde Joggen"; the media headline is from ISPO .com and Woman.at. The original study is Lee et al., "Running as a Key Lifestyle Medicine."

14. Reynolds, "An Hour of Running."

15. Tesla, "All Tesla Cars Being Produced."

16. Hawkins, "Here Are Elon Musk's Wildest Predictions."

17. Ewing, "German Court."

18. Topol, *Deep Medicine*.

19. This kind of argument to surrender to AI authority can, for instance, be found in Harari, "Big Data."

20. See Gigerenzer, *Risk Savvy*, chap. 5.

21. Lewis, *Moneyball*.

22. Lewis, *Moneyball*, 247.

23. See Hirsch and Hirsch, *Beauty of Short Hops*.

24. Gigerenzer et al., *Empire of Chance*.

25. Hirsch and Hirsch, *Beauty of Short Hops*, 32.

26. Barra, "Many Problems with 'Moneyball.'"

27. See, for instance, Harari, *Homo Deus*, 374.

28. Frederik and Martijn, "New Dot.com Bubble."

29. Frederik and Martijn, "New Dot.com Bubble."

30. Blake et al., "Consumer Heterogeneity."

31. Coviello et al., "Large-Scale Field Experiment."

32. An analysis of fifteen advertising experiments on Facebook, each with between 2 million and 140 million users, independently showed that correlational methods overestimate the effect of advertising. See Gordon et al., "Comparison of Approaches to Advertising Measurement."

33. Blake et al., "Consumer Heterogeneity." A related argument is that even if ads have no measurable direct effect, they may have a long-term effect on brand recognition. But companies with highly recognized brand names such as eBay spent most of their ad budget on customers already familiar with the brand who specifically searched for it, which makes brand-keyword advertising redundant.

34. Lewis and Rao, "On the Near Impossibility of Measuring the Returns."

35. Lewis and Reiley, "Advertising Effectively Influences Older Users."

36. See Hwang, *Subprime Attention Crisis.*

37. See Hwang, *Subprime Attention Crisis.*

38. Frederik and Martijn, "New Dot.com Bubble."

39. McGrew et al., "Improving University Students' Web Savvy."

40. McGrew et al., "Improving University Students' Web Savvy," 178.

41. https://www.minimumwage.com/2014/10/denmarks-dollar-forty-one-menu/.

42. Wineburg and McGrew, "Lateral Reading."

43. Wineburg and McGrew, "Lateral Reading."

44. McGrew et al., "Improving University Students' Web Savvy"; Breakstone et al., "Students' Civic Online Reasoning."

45. https://ccconline.libguides.com/webeval. See Kapoun, "Teaching Web Evaluation."

46. Neuvonen et al., *Elections Approach.* For the role of behavioral science in promoting techno-savvy, see Lorenz-Spreen et al., "How Behavioral Sciences Can Promote Truth," and Kozyreva et al., "Citizens Versus the Internet."

47. Weltecke, "Gab es 'Vertrauen' im Mittelalter?"

48. Mercier and Sperber, "Why Do Humans Reason?"

49. Howard, *Lie Machines.*

50. Rid, *Active Measures.*

51. Advisory Council for Consumer Affairs at the Federal Ministry of Justice and Consumer Protection, *Digital Sovereignty.*

52. Helbing et al., "Will Democracy Survive Big Data."

Bibliography

AddictionResource.net. "The Average Cost of Illegal Street Drugs." March 5, 2020. https://www.addictionresource.net/blog/cost-of-illegal-drugs/.

Advisory Council for Consumer Affairs at the Federal Ministry of Justice and Consumer Protection. *Consumer-Friendly Scoring: Recommendations for Action*. Berlin: Federal Ministry of Justice and Consumer Protection, 2018. https://www.svr-verbr aucherfragen.de/en/wp-content/uploads/sites/2/Recommandations-for-action.pdf.

Advisory Council for Consumer Affairs at the Federal Ministry of Justice and Consumer Protection. *Digital Sovereignty*. Berlin: Federal Ministry of Justice and Consumer Protection, 2017. https://www.svr-verbraucherfragen.de/wp-content/uploads /English-Version.pdf.

AFP. "'Shocking' Hack of Psychotherapy Records in Finland Affects Thousands." *Guardian*, October 26, 2020. https://www.theguardian.com/world/2020/oct/26/tens -of-thousands-psychotherapy-records-hacked-in-finland.

Aikman, D., M. Galesic, G. Gigerenzer, S. Kapadia, K. V. Katsikopoulos, A. Kothiyal, E. Murphy, and T. Neumann. "Taking Uncertainty Seriously: Simplicity Versus Complexity in Financial Regulation." *Industrial and Corporate Change* 30 (2021): 317–345.

Allen, M. "Sean Parker Unloads on Facebook: 'God Only Knows What It's Doing to Our Children's Brains.'" *Axios*, November 9, 2017. https://www.axios.com/sean -parker-unloads-on-facebook-god-only-knows-what-its-doing-to-our-childrens-brains -1513306792-f855e7b4-4e99-4d60-8d51-2775559c2671.html.

Anderson, C. "The End of Theory: The Data Deluge Makes the Scientific Method Obsolete." *Wired*, June 23, 2008. https://www.wired.com/2008/06/pb-theory/.

Anderson, G. "What Happens to Big Data If the Small Data Is Wrong?" *RetailWire*, September 10, 2013. https://retailwire.com/discussion/what-happens-to-big-data-if -the-small-data-is-wrong/.

Anderson, M., E. A. Vogels, and E. Turner. "The Upsides and Downsides of Online Dating." Pew Research Center, February 6, 2020. https://www.pewresearch.org /internet/2020/02/06/the-virtues-and-downsides-of-online-dating/.

Anderson, R. "The Ugly Truth about Dating: Are We Sacrificing Love for Convenience?" *Psychology Today*, September 6, 2016. https://www.psychologytoday.com /us/blog/the-mating-game/201609/the-ugly-truth-about-online-dating.

Angelino, E., N. Larus-Stone, D. Alabi, M. Seltzer, and C. Rudin. "Learning Certifiably Optimal Rule Lists for Categorial Data." *Journal of Machine Learning Research* 18 (2018): 1–78.

Armstrong, J. S. *Persuasive Advertising: Evidence-Based Principles*. London: Palgrave Macmillan, 2010.

Artinger, F. M., N. Kozodi, F. Wangenheim, and G. Gigerenzer. "Recency: Prediction with Smart Data." In *2018 AMA Winter Academic Conference: Integrating Paradigms in a World Where Marketing Is Everywhere*, AMA Educators Proceedings, vol. 29, edited by J. Goldenberg, J. Laran, and A. Stephen, L-2–6. Chicago: American Marketing Association, 2018.

Asch, S. E. "Effects of Group Pressure on the Modification and Distortion of Judgments." In *Groups, Leadership, and Men*, edited by H. Guetzkow, 177–190. Pittsburgh: Carnegie Press, 1951.

Aschwanden, C. "You Can't Trust What You Read about Nutrition." *FiveThirtyEight*, January 6, 2016. https://fivethirtyeight.com/features/you-cant-trust-what-you-read -about-nutrition/.

Awad, E., S. Dsouza, R. Kim, J. Schulz, J. Heinrich, A. Shariff, J.-F. Bonnefon, and I. Rahwan. "The Moral Machine Experiment." *Nature* 563 (2018): 59–64.

Bailey, D. H., J. M. Borwein, M. Lopez de Prado, and Q. Zhu. "Pseudo-Mathematics and Financial Charlatanism: The Effect of Backtest Overfitting on Out-of-Sample Prediction." *Notices of the American Mathematical Society* 61 (2014): 458–471.

Baltrusaitis, J. "Top 10 Countries and Cities by Number of CCTV Cameras." *Precise Security*, December 4, 2019 (updated June 20, 2020). https://www.precisesecurity .com/articles/Top-10-Countries-by-Number-of-CCTV-Cameras.

Baraniuk, C. "GPS Error Caused '12 Hours of Problems' for Companies." *BBC News*, February 4, 2016. https://www.bbc.com/news/technology-35491962.

Barnes, S. B. "A Privacy Paradox: Social Networking in the United States." *First Monday* 11, no. 9 (2006). https://firstmonday.org/article/view/1394/.

Barra, A. "The Many Problems with 'Moneyball.'" *Atlantic*, September 27, 2011. https://www.theatlantic.com/entertainment/archive/2011/09/the-many-problems -with-moneyball/245769/.

Barrowman, N. "Correlation, Causation, and Confusion." *The New Atlantis*, 2014. https://www.thenewatlantis.com/publications/correlation-causation-and-confusion.

Begley, S. "In Cancer Science, Many 'Discoveries' Don't Hold Up." *Science News*, March 28, 2012. https://www.reuters.com/article/us-science-cancer-%20idUSBRE82 R12P20120328.

Bennell, C., P. J. Taylor, and B. Snook. "Clinical Versus Actuarial Geographic Profiling Strategies: A Review of the Research." *Police Practice & Research* 8 (2007): 335–345.

Benoliel, U., and S. I. Becher. "The Duty to Read the Unreadable." *Boston College Law Review* 60 (2019): 2255–2296.

Berg, N., and G. Gigerenzer. "As-if Behavioral Economics: Neoclassical Economics in Disguise?" *History of Economic Ideas* 18 (2010): 133–165.

Bergin, T. "How a Data Mining Giant Got Me Wrong." *Reuters*, March 29, 2018. https://www.reuters.com/article/us-data-privacy-acxiom-insight/how-a-data-mining -giant-got-me-wrong-idUSKBN1H513K.-

Berwick, A. "How ZTE Helps Venezuela Create China-Style Social Control." *Reuters*, November 14, 2018. https://www.reuters.com/investigates/special-report/venezuela -zte/.

Best, J. "IBM Watson: The Inside Story of How the Jeopardy-Winning Supercomputer Was Born, and What It Wants to Do Next." *TechRepublic*, September 9, 2013. https://www.techrepublic.com/article/ibm-watson-the-inside-story-of-how-the -jeopardy-winning-supercomputer-was-born-and-what-it-wants-to-do-next/.

Blake, T., C. Nosko, and S. Tadelis. "Consumer Heterogeneity and Paid Search Effectiveness: A Large Scale Field Experiment." *Econometrica* 83 (2015): 155–174.

Bliss, L. "How Utrecht Became a Paradise for Cyclists." *Bloomberg CityLab*, July 5, 2019. https://www.bloomberg.com/news/articles/2019–07–05/how-the-dutch-made -utrecht-a-bicycle-first-city.

Boden, M. *Mind as Machine: A History of Cognitive Science*. Oxford: Oxford University Press, 2008.

Bond, R. M., C. J. Fariss, J. J. Jones, A. D. I. Kramer, C. Marlow, J. E. Settle, and J. H. Fowler. "A 61-Million-Person Experiment in Social Influence and Political Mobilization." *Nature* 489 (2012): 295–298.

Booth, R. "Facebook Reveals News Feed Experiment to Control Emotions." *Guardian*, June 30, 2014. https://www.theguardian.com/technology/2014/jun/29/facebook-users -emotions-news-feeds.

Bostrom, N. *Superintelligence*. Oxford: Oxford University Press, 2015.

Brainbridge, L. "Ironies of Automation." *Automatica* 19 (1983): 775–779.

Brandstätter, E., G. Gigerenzer, and R. Hertwig. "Risky Choice with Heuristics: Reply to Birnbaum (2008), Johnson, Schulte-Mecklenbeck, and Willemsen (2008), and Rieger and Wang (2008)." *Psychological Review* 115, no. 1 (2008): 281–289.

Branwen, G. "The Neural Net Tank Urban Legend." Blog post, 2011. https://www .gwern.net/Tanks.

Breakstone, J., M. Smith, P. Connors, T. Ortega, D. Kerr, and S. Wineburg. "Lateral Reading: College Students Learn to Critically Evaluate Internet Sources in an Online Course." *Harvard Kennedy School (HKS) Misinformation Review* (2021). https://doi .org/10.37016/mr-2020-56.

Breakstone, J., M. Smith, S. Wineburg, A. Rapaport, J. Carle, M. Garland, and A. Saavedra. "Students' Civic Online Reasoning: A National Portrait." *Educational Researcher* (May 2021). https://doi.org/10.3102/0013189X211017495.

Breiman, L. "Statistical Modeling: The Two Cultures (with Comments and a Rejoinder by the Author)." *Statistical Science* 16 (2001): 199–231.

Brin, S., and L. Page. "Reprint of: The Anatomy of a Large-Scale Hypertextual Web Search Engine." *Computer Networks* 56 (2012): 3825–3833.

Britt, R. R. "Drivers on Cell Phones Kill Thousands, Snarl Traffic." *Live Science*, February 1, 2015. https://www.livescience.com/121-drivers-cell-phones-kill-thousands-snarl -traffic.html.

Brockman, J., ed. *Possible Minds*. New York: Penguin, 2019.

Brooks, R. "Mistaking Performance for Competence." In *What to Think about Machines That Think*, edited by J. Brockman, 108–111. New York: Harper, 2015.

Brown, A. "Couples Who Meet Online Are More Diverse Than Those Who Meet in Other Ways, Largely Because They're Younger." Pew Research Center, June 24, 2019. https://www.pewresearch.org/fact-tank/2019/06/24/couples-who-meet-online-are -more-diverse-than-those-who-meet-in-other-ways-largely-because-theyre-younger/.

Brown, J. "Why Everyone Is Hating on IBM Watson—Including the People Who Helped Make It." *Gizmodo*, August 20, 2017. https://gizmodo.com/why-everyone-is -hating-on-watson-including-the-people-w-1797510888.

Brown, K. V. "We Talked to 24 Victims of the Ashley Madison Hack about Their Exposed Secrets." *Splinter*, August 19, 2015. https://splinternews.com/we-talked-to -24-victims-of-the-ashley-madison-abou-1793850144.

Brown, T. *Lectures on the Philosophy of the Human Mind*. London: William Tait, 1838.

Bruch, E. E., and M. E. J. Newman. "Aspirational Pursuit of Mates in Online Dating Markets." *Science Advances* 4, no. 8 (2018): eaap9815.

Brumfiel, G. "U.S. Navy Brings Back Navigation by the Stars for Officers." *NPR*, February 28, 2016. https://text.npr.org/467210492.

Brunswik, E. *Perception and the Representative Design of Psychological Experiments*. Los Angeles: University of California Press, 1956.

Buolamwini, J., and T. Gebru. "Gender Shades: Intersectional Accuracy Disparities in Commercial Gender Classification." *Proceedings in Machine Learning Research* 81 (2018): 1–15.

Burell, J. "How the Machine 'Thinks': Understanding Opacity in Machine Learning Algorithms." *Big Data & Society* (January–June 2016): 1–12.

Buss, D. *Evolutionary Psychology: The New Science of the Mind*. 6th ed. New York: Routledge, 2019.

Cacioppo, J., S. Cacioppo, G. C. Gonzaga, E. L. Ogburn, and T. J. VanderWeele. "Marital Satisfaction and Break-Ups Differ Across On-Line and Off-Line Meeting Venues." *PNAS* 110 (2013): 10135–10140.

Cahall, B., P. Bergen, D. Sterman, and E. Schneider. "Do NSA's Bulk Surveillance Programs Stop Terrorists?" *New America*, January 13, 2014. https://www.newamerica .org/international-security/policy-papers/do-nsas-bulk-surveillance-programs-stop -terrorists/.

Calaprice, A. *The Ultimate Quotable Einstein*. Princeton, NJ: Princeton University Press, 2011.

Carr, N. *The Glass Cage*. New York: Norton, 2014.

Carroll, L. *The Hunting of the Snark*. London: Macmillan and Co., 1876. Reprint, Copenhagen: SAGA Egmont, 2020.

Centers for Disease Control and Prevention. "Transportation Safety: Distracted Driving." October 6, 2020. https://www.cdc.gov/transportationsafety/distracted_driving/.

Chalmers, I., and P. Glasziou. "Avoidable Waste in the Production and Reporting of Research Evidence." *Lancet* 374, no. 9683 (2009): 86–89.

Chawla, D. S. "The Need for Digital Intelligence." *Nature* 562 (2018): S15–S16.

Chin, C. "AI Is the Future—But Where Are the Women?" *Wired*, August 17, 2018. https://www.wired.com/story/artificial-intelligence-researchers-gender-imbalance/.

Chiusi, F. "Life in the Automated Society: How Automated Decision-Making Systems Became Mainstream, and What to Do about It." In *Automating Society Report 2020*. AlgorithmWatch/Bertelsmann Stiftung, 2020. https://automatingsociety.algorithm watch.org/.

Cho, E. "The Social Credit System: Not Just Another Chinese Idiosyncrasy." *Journal of Public and International Affairs*, May 1, 2020. https://jpia.princeton.edu/news/social-credit-system-not-just-another-chinese-idiosyncrasy.

Christl, W. *Corporate Surveillance in Everyday Life: How Companies Collect, Combine, Analyze, Trade, and Use Personal Data on Billions*. Vienna: Cracked Labs, 2017. https://crackedlabs.org/dl/CrackedLabs_Christl_CorporateSurveillance.pdf.

Church, G. M. "The Rights of Machines." In *Possible Minds*, edited by J. Brockman, 240–253. New York: Penguin, 2019.

Clayton, V., M. Sanders, E. Schoenwald, L. Surkis, and D. Gibbons. *Machine Learning in Children's Services*. What Works for Children's Social Care, September 2020. https://whatworks-csc.org.uk/wp-content/uploads/WWCSC_technical-_report_machine_learning_in_childrens_services_does_it_work_Sep_2020.pdf.

Clifford, S., L. D. Doane, R. Breitenstein, K. J. Grimm, and K. Lemery-Chalfant. "Effortful Control Moderates the Relation Between Electronic-Media Use and Objective Sleep Indicators in Childhood." *Psychological Science* 31 (2020): 822–834.

CNS/ATM. *CNS/ATM Resource Guide* (chapter 8). Canberra: Civil Aviation Safety Authority, 2017. https://www.casa.gov.au/book-page/chapter-8-human-factors.

Cohen, I. B. "Howard Aiken on the Number of Computers Needed for the Nation." *IEEE Annals of the History of Computing* 20 (1998): 27–32.

Coleridge, S. T. *The Literary Remains of Samuel Taylor Coleridge*. Vol. 3. London: H. Pickering, 1838. Accessed April 5, 2021. https://archive.org/details/literaryremainso03coleuoft/page/186/mode/2up.

Connolly, K. "Child Drowning in Germany Linked to Parents' Phone 'Fixation.'" *Guardian*, August 15, 2018. https://www.theguardian.com/lifeandstyle/2018/aug/15/parents-fixated-by-phones-linked-to-child-drownings-in-germany.

Copeland, P., R. Romano, T. Zhang, G. Hecht, D. Zigmond, and C. Stefansen. "Google Disease Trends: An Update." https://storage.googleapis.com/pub-tools-public-publication-data/pdf/41763.pdf.

Coppock, A., S. J. Hill, and L. Vavreck. "The Small Effects of Political Advertising Are Small Regardless of Context, Message, Sender, or Receiver: Evidence From 59 Real-Time Randomized Experiments." *Science Advances* 6 (2020): eabc4046.

Coviello, L., U. Gneezy, and L. Götte. *A Large-Scale Field Experiment to Evaluate the Effectiveness of Paid Search Advertising*. CESifo Working paper 6684, Center for Economic Studies and ifo Institute, Munich, 2017.

Czerlinski, J., G. Gigerenzer, and D. G. Goldstein. "How Good Are Simple Heuristics?" In *Simple Heuristics That Make Us Smart*, by G. Gigerenzer, P. M. Todd, and the ABC Research Group, 97–118. New York: Oxford University Press, 1999.

Daniel, C., and M. Palmer. "Google's Goal: To Organise Your Daily Life." *Financial Times*, May 27, 2007. https://www.ft.com/content/c3e49548-088e-11dc-b11e-000b 5df10621.

Danielsbacka, M., A. O. Tanskanen, and F. C. Billari. "Meeting Online and Family-Related Outcomes: Evidence from Three German Cohorts." *Journal of Family Studies* (2020). https://doi.org/10.1080/13229400.2020.1835694.

Danziger, S., J. Levav, and L. Avnaim-Pesso. "Extraneous Factors in Judicial Decisions." *PNAS* 108 (2011): 6889–6892.

Dastin, J. "Amazon Scraps Secret AI Recruiting Tool That Showed Bias against Women." *Reuters*, October 11, 2018. https://www.reuters.com/article/us-amazon -com-jobs-automation-insight-idUSKCN1MK08G.

Daston, L. "Calculation and the Division of Labor, 1750–1950." *Bulletin of the German Historical Institute* 62 (2018): 9–30.

Daston, L. "Enlightenment Calculations." *Critical Inquiry* 21 (1994): 182–202.

Daston, L. "The Immortal Archive: Nineteenth-Century Science Imagines the Future." In *Science in the Archives: Pasts, Presents, Futures*, edited by L. Daston, 159–182. Chicago: University of Chicago Press, 2017.

Daston, L. *Rules: A Short History of What We Live By*. Princeton, NJ: Princeton University Press, 2022.

Davis, B. D., J. C. Mason, and M. Anwar. "Vulnerability Studies and Security Postures of IoT Devices: A Smart Home Case Study." *IEEE Internet of Things Journal* 7 (2020): 10102–10110.

Dawes, R., and B. Corrigan. "Linear Models in Decision Making." *Psychological Bulletin* 81, no. 2 (1974): 95–106.

DeMichele, M., P. Baumgartner, M. Wenger, K. Barrick, and M. Comfort. "Public Safety Assessment: Predictive Utility and Differential Prediction by Race in Kentucky." *Criminology & Public Policy* 19 (2020): 409–431.

Deutsches Institut für Service-Qualität. "Kundenbefragung Online-Partnerbörsen" [Survey on Dating Services]. 2017. https://disq.de/2017/20170426-Online-Partner boersen.html#GesamtP.

Dickmanns, E. D., and A. Zapp. "Autonomous High Speed Road Vehicle Guidance by Computer Vision." *IFAC Proceedings Volumes* 20 (1987): 221–226.

Digital Courage. "'Stasi-Barbie'—Der Spion in der Spielzeugkiste" ["Stasi Barbie": The Spy in the Toybox]. Press release, January 26, 2016. https://digitalcourage.de /blog/2016/stasi-barbie-der-spion-in-der-spielzeugkiste.

Dillet, R. "Renault-Nissan CEO Carlos Ghosn on the Future of Cars." *TechCrunch*, October 13, 2016. https://techcrunch.com/2016/10/13/renault-nissan-ceo-carlos-ghosn-on-the-future-of-cars/.

Dingus, T. A., F. Guo, S. Lee, J. F. Antin, M. Perez, M. Buchanan-King, and J. Hankey. "Driver Crash Risk Factors and Prevalence Evaluation Using Naturalistic Driving Data." *Proceedings of the National Academy of Sciences* 113, no. 10 (2016): 2636–2341.

Dittmar, H., E. Halliwell, and S. Ive. "Does Barbie Make Girls Want to Be Thin? The Effect of Experimental Exposure to Images of Dolls on the Body Image of 5- to 8-Year-Old Girls." *Developmental Psychology* 42, no. 2 (2006): 283–292.

Doll, N. "Moderne Autos Sind Datenkraken" [Modern Cars Are Data Leeches]. *Welt*, November 22, 2016. https://www.welt.de/print/die_welt/finanzen/article159 665896.

Dosi, G., M. Napoletano, A. Roventini, J. E. Stiglitz, and T. Treibich. "Rational Heuristics? Expectations and Behaviors in Evolving Economies with Heterogeneous Interacting Agents." *Economic Inquiry* 53 (2020): 1487–1516.

Dressel, J., and H. Farid. "The Accuracy, Fairness, and Limits of Predicting Recidivism." *Science Advances* 4, no. 1 (2018): eaao5580.

Dreyfus, H. L. *What Computers Can't Do*. Rev. ed. New York: Harper, 1979.

Dyrenforth, P. S., D. A. Kashy, M. B. Donnellan, and R. E. Lucas. "Predicting Relationship and Life Satisfaction from Personality in Nationally Representative Samples from Three Countries: The Relative Importance of Actor, Partner, and Similarity Effects." *Journal of Personality and Social Psychology* 99 (2010): 690–702.

Eames, S. S. *Barbie Doll Fashion: 1959–1967*. Paducah, KY: Collector Books, 1990.

Efrati, A. "How an Uber Whistleblower Tried to Stop Self-Driving Car Disaster." *The Information*, December 10, 2018. https://www.theinformation.com/articles/how-an-uber-whistleblower-tried-to-stop-self-driving-car-disaster.

Eggers, D. *The Circle*. New York: Knopf, 2013.

Einhorn, H. J., and R. Hogarth. "Unit Weighting Schemes for Decision Making." *Organizational Behavior and Human Performance* 13, no. 2 (1975): 171–192.

Elias, J. "Alphabet Exec Says Self-Driving Cars 'Have Gone through a Lot of Hype,' but Google Helped Drive That Hype." *CNBC*, October 23, 2019. https://www.cnbc.com/2019/10/23/alphabet-exec-admits-google-overhyped-self-driving-cars.html.

Elliott, J., and T. Meyer. "Claim on 'Attacks Thwarted' by NSA Spreads Despite Lack of Evidence." *ProPublica*, October 23, 2013. https://www.propublica.org/article/claim-on-attacks-thwarted-by-nsa-spreads-despite-lack-of-evidence.

Elsberg, M. *Blackout*. London: Penguin Books, 2017.

Epstein, R. *Simple Rules for a Complex World*. Cambridge, MA: Harvard University Press, 1995.

Epstein, R. "The Truth about Online Dating." In *Disarming Cupid: Love, Sex, and Science*, edited by Editors of Scientific American. New York: Scientific American, 2013.

Epstein, R., and R. E. Robertson. "The Search Engine Manipulation Effect (SEME) and Its Possible Impact on the Outcomes of Elections." *PNAS* 112, no. 33 (2015): E4512–E4521.

ERGO. *ERGO Risiko-Report* [ERGO Risk Report]. 2019. Accessed April 5, 2021. https://www.ergo.com/en/Media-Relations/Pressemeldungen/PM-2019/20190912-ERGO-Risiko-Report.

European Commission. *Artificial Intelligence—A European Approach to Excellence and Trust*. 2020. https://ec.europa.eu/info/sites/info/files/commission-white-paper-artificial-intelligence-feb2020_en.pdf.

European Union. *General Data Protection Regulation*. 2016. https://gdpr-info.eu/art-5-gdpr/.

Ewing, J. "German Court Says Tesla Self-Driving Claims Are Misleading." *New York Times*, July 14, 2020. https://www.nytimes.com/2020/07/14/business/tesla-autopilot-germany.html.

Eyal, N. *Indistractable: How to Control Your Attention and Choose Your Life*. London: Bloomsbury, 2019.

Eyal, N. *Hooked*. London: Penguin, 2014.

Facebook. "Facebook Reports Fourth Quarter and Full Year 2019 Results." Press release, January 29, 2020. https://investor.fb.com/investor-news/press-release-details/2020/Facebook-Reports-Fourth-Quarter-and-Full-Year-2019-Results/default.aspx.

Farnham, A. "Hot or Not's Co-Founders: Where Are They Now?" *ABC News*, June 2, 2014. https://abcnews.go.com/Business/founders-hot-today/story?id=23901082.

Federal Aviation Administration. "Safety Alert for Operators: Manual Flight Operations (SAFO 13002)." January 4, 2013. https://www.faa.gov/other_visit/aviation_industry/airline_operators/airline_safety/safo/all_safos/media/2013/SAFO13002.pdf.

Federal Trade Commission. "FTC Sues Owner of Online Dating Service Match.Com for Using Fake Love Interest Ads to Trick Consumers into Paying for a Match.Com Subscription." Press release, September 25, 2019. https://www.ftc.gov/news-events/press-releases/2019/09/ftc-sues-owner-online-dating-service-matchcom-using-fake-love.

Ferrucci, D., E. Brown, J. Chu-Carroll, J. Fan, D. Gondek, A. A. Kalyanpur, A. Lally, et al. "Building Watson: An Overview of the DeepQA Project." *AI Magazine* 31, no. 3 (2010): 59–79.

Feynman, R. P., R. Leighton, and E. Hutchings. *"Surely You're Joking, Mr. Feynman!":* *Adventures of a Curious Character.* New York: W. W. Norton, 1985.

Finkel, E. J., P. W. Eastwick, B. R. Karney, H. T. Reis, and S. Sprecher. "Online Dating: A Critical Analysis from the Perspective of Psychological Science." *Psychological Science in the Public Interest* 13 (2012): 3–66.

Fleischhut, N., B. Meder, and G. Gigerenzer. "Moral Hindsight." *Experimental Psychology* 64 (2017): 110–123.

Fletcher, D. "The 50 Worst Inventions: Farmville." *Time*, May 27, 2010. http://content .time.com/time/specials/packages/article/0,28804,1991915_1991909_1991768,00 .html.

"Flying to Distraction." *Flight Safety Australia*, July 6, 2018. https://www.flightsafety australia.com/2018/07/flying-to-distraction/.

Fowler, G. A. "What Does Your Car Know about You? We Hacked a Chevy to Find Out." *Washington Post*, December 27, 2019. https://www.washingtonpost.com /technology/2019/12/17/what-does-your-car-know-about-you-we-hacked-chevy -find-out/.

Fowler, G. A. "You Watch TV. Your TV Watches Back." *Washington Post*, September 18, 2019. https://www.washingtonpost.com/technology/2019/09/18/you-watch-tv -your-tv-watches-back/.

Frederik, J., and M. Martin. "The New Dot.com Bubble Is Here: It's Online Advertising." *The Correspondent*, November 6, 2019. https://thecorrespondent.com/100/the -new-dot-com-bubble-is-here-its-called-online-advertising/13228924500–22d5fd24.

Freed, D., J. Palmer, D. Minchala, K. Levy, T. Ristenpart, and N. Dell. "'A Stalker's Paradise': How Intimate Partner Abusers Exploit Technology." In *CHI '18*. New York: ACM Press, 2018. http://nixdell.com/papers/stalkers-paradise-intimate.pdf.

Geirhos, R., J.-H. Jacobsen, C. Michaelis, R. Zemel, W. Brendel, M. Bethge, and F. A. Wichmann. "Shortcut Learning in Deep Neural Networks." *Nature Machine Intelligence* 2 (2020): 665–673.

Gendron, S. "Croisade Contre les Textos au Volant" [A Crusade against Texting while Driving]. *Le Journal de Montréal*, May 4, 2016. https://www.journaldemontreal .com/2016/05/04/croisade-contre-les-textos-au-volant.

Geraghty, K., and J. Woodhams. "The Predictive Validity of Risk Assessment Tools for Female Offenders: A Systematic Review." *Aggression and Violent Behavior* 21 (2015): 25–38.

Gibbs, S. "Apple's Tim Cook: 'I Don't Want My Nephew on a Social Network.'" *Guardian*, January 19, 2018. https://www.theguardian.com/technology/2018/jan/19 /tim-cook-i-dont-want-my-nephew-on-a-social-network.

Gigerenzer, G. *Calculated Risks: How to Know When Numbers Deceive You.* New York: Simon & Schuster, 2002.

Gigerenzer, G. "Digital Computer: Impact on the Social Sciences." In *International Encyclopedia of the Social and Behavioral Sciences*, vol. 6, edited by N. J. Smelser and P. B. Baltes, 3684–3688. Amsterdam: Elsevier, 2001.

Gigerenzer, G. "The Frequency-Validity Relationship." *American Journal of Psychology* 97 (1984): 185–195.

Gigerenzer, G. "From Tools to Theories: A Heuristic of Discovery in Cognitive Psychology." *Psychological Review* 98 (1991): 254–267.

Gigerenzer, G. *Gut Feelings: The Intelligence of the Unconscious.* New York: Viking, 2007.

Gigerenzer, G. *Risk Savvy: How to Make Good Decisions.* New York: Viking, 2014.

Gigerenzer, G. "What Is Bounded Rationality?" In *Routledge Handbook of Bounded Rationality*, edited by R. Viale, 55–69. London: Routledge, 2021.

Gigerenzer, G., W. Gaissmaier, E. Kurz-Milcke, L. M. Schwartz, and S. W. Woloshin. "Helping Doctors and Patients Make Sense of Health Statistics." *Psychological Science in the Public Interest* 8 (2007): 53–96.

Gigerenzer, G., and D. G. Goldstein. "Mind as Computer: Birth of a Metaphor." *Creativity Research Journal* 9 (1996): 131–144.

Gigerenzer, G., R. Hertwig, and T. Pachur, eds. *Heuristics: The Foundations of Adaptive Behavior.* New York: Oxford University Press, 2011.

Gigerenzer, G., W. Krämer, and T. K. Bauer. "Eine Stunde Joggen, Sieben Stunden Länger Leben" [One Hour of Jogging Adds Seven Hours to Your Life]. *Unstatistik*, 28 April, 2017. https://www.rwi-essen.de/unstatistik/66/.

Gigerenzer, G., and J. A. Muir Gray, eds. *Better Doctors, Better Patients, Better Decisions.* Cambridge, MA: MIT Press, 2011.

Gigerenzer, G., and D. J. Murray. *Cognition as Intuitive Statistics.* Mahwah, NJ: Erlbaum, 1987. Reprint, New York: Psychology Press, 2015.

Gigerenzer, G., Z. Swijtink, T. Porter, L. Daston, J. Beatty, and L. Kruger. *The Empire of Chance: How Probability Changed Science and Everyday Life.* Cambridge: Cambridge University Press, 1989.

Ginsberg, J., M. H. Mohebbi, R. S. Patel, L. Brammer, M. S. Smolinski, and L. Brilliant. "Detecting Influenza Epidemics Using Search Engine Query Data." *Nature* 457 (2009): 1012–1014.

Gleave, A. "Physically Realistic Attacks on Deep Reinforcement Learning." Berkeley Artificial Intelligence Research blog, March 27, 2020. https://bair.berkeley.edu/blog/2020/03/27/attacks/.

Gleick, J. *Genius: The Life and Science of Richard Feynman.* New York: Pantheon, 1992.

Gliklich, J., R. Maurer, and R. W. Bergmark. "Patterns of Texting and Driving in a US National Survey of Millennial Parents vs Older Parents." *JAMA Pediatrics* 173 (2019): 689–690.

Glöckner, A. "The Irrational Hungry Judge Effect Revisited: Simulations Reveal that the Magnitude of the Effect Is Overestimated." *Judgment and Decision Making* 11 (2016): 601–610.

Gordon, B. R., F. Zettelmeyer, N. Bhargave, and D. Chapsky. "A Comparison of Approaches to Advertising Measurement: Evidence from Big Field Experiments at Facebook." *Marketing Science* 38 (2019): 193–364.

Gorner, J., and A. Sweeney. "For Years Chicago Police Rated the Risk of Tens of Thousands Being Caught Up in Violence. That Controversial Effort Has Quietly Been Ended." *Chicago Tribune*, January 24, 2020. https://www.chicagotribune.com/news/criminal-justice/ct-chicago-police-strategic-subject-list-ended-20200125-spn4kjmrxrh4tmktdjckhtox4i-story.html.

Gottman, J., and J. Gottman. "The Natural Principles of Love." *Journal of Family Theory & Review* 9 (2017): 7–26.

Gottman, J. M., J. Coan, S. Carrere, and C. Swanson. "Predicting Marital Happiness and Stability from Newlywed Interactions." *Journal of Marriage and the Family* 60 (1998): 5–22.

Götz, M., E. Wunderer, J. Greithanner, and E. Maslanka. "'Warum Kann Ich Nicht so Perfekt Sein?'" [Why Can't I Be as Perfect?]. *Televizion* 32, no. 1 (2019). https://www.br-online.de/jugend/izi/deutsch/publikation/televizion/32_2019_1/Goetz_Wunderer-perfekt_sein.pdf.

Green, K. C., and J. S. Armstrong. "Simple Versus Complex Forecasting: The Evidence." *Journal of Business Research* 68, no. 1 (2015): 678–685.

Greenwald, G. *No Place to Hide: Edward Snowden, the NSA and the Surveillance State.* London: Penguin Books, 2014.

Griffiths, J. *The Great Firewall of China.* London: Zed Books, 2019.

Grisso, T., and A. J. Tomkins. "Communicating Violence Risk Assessments." *American Psychologist* 51, no. 9 (1996): 928–930.

Grossman, L., M. Thompson, J. Kluger, A. Park, B. Walsch, C. Suddath, E. Dodds, et al. "The 50 Best Inventions." *Time*, November 28, 2011. http://content.time.com/time/subscriber/article/0,33009,2099708-1,00.html.

Grzymek, V., and M. Puntschuh. *What Europe Knows and Thinks about Algorithms: Results of a Representative Survey*. Gütersloh: Bertelsmann Stiftung, 2019. https://www.bertelsmann-stiftung.de/fileadmin/files/BSt/Publikationen/GrauePublikationen/WhatEuropeKnowsAndThinkAboutAlgorithm.pdf.

Gunning, D., and D. W. Aha. "DARPA's Explainable Artificial Intelligence Program." *AI Magazine* 40, no. 2 (2019): 44–58.

Gururangan, S., S. Swayamdipta, O. Levy, R. Schwartz, S. Bowman, and N. A. Smith. "Annotation Artifacts in Natural Language Inference Data." *Proceedings of the 2018 Conference of the North American Chapter of the Association for Computational Linguistics: Human Language Technologies 2* (2018): 107–112. https://www.aclweb.org/anthology/N18-2017/.

Halpern, S. "The Creepy New Wave of the Internet." *New York Review of Books*, November 20, 2014, 22–24.

Hancock, J. T., C. Toma, and N. Ellison. "The Truth about Lying in Online Dating Profiles." *CHI 2007 Proceedings*. San Jose, CA: CHI, 2007. https://collablab.northwestern.edu//CollabolabDistro/nucmc/p449-hancock.pdf.

Hao, K. "We Read the Paper That Forced Timnit Gebru Out of Google. Here's What It Says." *MIT Technology Review*, December 4, 2020. https://www.technologyreview.com/2020/12/04/1013294/google-ai-ethics-research-paper-forced-out-timnit-gebru/.

Harari, Y. N. *Homo Deus: A Brief History of Tomorrow*. New York: Harper, 2015.

Harari, Y. N. "Yuval Noah Harari on Big Data, Google and the End of Free Will." *Financial Times*, August 16, 2016. https://www.ft.com/content/50bb4830–6a4c-11e6-ae5b-a7cc5dd5a28c.

Harford, T. "Big Data: Are We Making a Big Mistake?" *Financial Times*, March 28, 2014. https://www.ft.com/content/21a6e7d8-b479–11e3-a09a-00144feabdc0.

Hartmann, S. "Risikobewältigung in der Luftfahrt und in der Medizin—Eine Vergleichende Untersuchung" [Managing Risk in Air Traffic and Medicine: A Comparative Analysis]. PhD diss., Humboldt University Berlin, 2017.

Hauber, J., E. Jarchow, and S. Rabitz-Suhr. *Prädiktionspotenzial Schwere Einbruchskriminalität: Ergebnisse einer Wissenschaftlichen Befassung mit Predictive Policing* [The Potential of Predicting Serious Burglary Offences: Results of a Scientific Investigation on Predictive Policing]. Hamburg: Landeskriminalamt Hamburg, 2019. https://www.polizei.hamburg/contentblob/13755082/74aff9285c340ad260ed65fe912caa17/data/abschlussbericht-praediktionspotenzial-schwere-einbruchskrminalitaet-do.pdf.

Hawkins, A. J. "Here Are Elon Musk's Wildest Predictions about Tesla's Self-Driving Cars." *The Verge*, April 22, 2019. https://www.theverge.com/2019/4/22/18510828/tesla-elon-musk-autonomy-day-investor-comments-self-driving-cars-predictions.

Hayes, B. "Gauss's Day of Reckoning." *American Scientist* 94 (2006): 200–205.

Haynes, T. "Dopamine, Smartphones, and You: A Battle for Your Time." Harvard University, May 1, 2018. http://sitn.hms.harvard.edu/flash/2018/dopamine-smart phones-battle-time/.

Helbing, D., B. S. Frey, G. Gigerenzer, E. Hafen, M. Hagner, Y. Hofstetter, J. van den Hoven, R. V. Zicari, and A. Zwitter. "Will Democracy Survive Big Data and Artificial Intelligence?" *Scientific American*, February 25, 2017. https://www.scientificamerican .com/article/will-democracy-survive-big-data-and-artificial-intelligence/.

Heyman, R. E., and A. M. S. Slep. "The Hazards of Predicting Divorce without Cross-validation." *Journal of Marriage and the Family* 63 (2001): 473–479.

Hirsch, S., and A. Hirsch. *The Beauty of Short Hops: How Chance and Circumstance Confound the Moneyball Approach to Baseball*. Jefferson, NC: McFarland, 2011.

Hitsch, G. J., A. Hortaçsu, and D. Ariely. "Matching and Sorting in Online Dating." *American Economic Review* 100 (2010): 130–163.

Hofstadter, D. "The Shallowness of Google Translate." *Atlantic*, January 30, 2018. https://www.theatlantic.com/technology/archive/2018/01/the-shallowness-of -google-translate/551570/.

Holte, C. H. "Very Simple Classification Rules Perform Well on Most Commonly Used Datasets." *Machine Learning* 11 (1993): 63–91.

Holzinger, A., C. Biemann, C. S., Pattichis, and D. B. Kell. "What Do We Need to Build Explainable AI Systems for the Medical Domain?" *ArXiv*, December 28, 2017. https://arxiv.org/abs/1712.09923.

Howard, P. N. *Lie Machines*. New Haven, CT: Yale University Press, 2020.

Hsu, J. W. "Transasia Pilot Shut Down Wrong Engine Before Taiwan Crash." *Wall Street Journal*, July 2, 2015. https://www.wsj.com/articles/transasia-pilot-shut-down -working-engine-before-taiwan-crash-1435815552.

Hu, J. C. "How One Light Bulb Could Allow Hackers to Burgle Your Home." *Quartz*, December 18, 2018. https://qz.com/1493748/how-one-lightbulb-could-allow-hackers -to-burgle-your-home/.

Huang, S., S. Aral, Y. J. Hu, and E. Brynjolfsson. "Social Advertising Effectiveness Across Products: A Large-Scale Field Experiment." *Marketing Science* 39(2020): 1142–1165.

Hwang, T. *Subprime Attention Crisis: Advertising and the Time Bomb at the Heart of the Internet*. New York: Macmillan, 2020.

Instituto Superiore di Sanità. *Characteristics of COVID-19 Patients Dying in Italy*. Report based on available data from November 18, 2020. https://www.epicentro.iss .it/en/coronavirus/sars-cov-2-analysis-of-deaths.

Isaac, M. "Facebook Said to Create Censorship Tool to Get Back into China." *New York Times*, November 22, 2016. https://www.nytimes.com/2016/11/22/technology /facebook-censorship-tool-china.html.

Islam, M. S., T. Sarkar, Hossain S. Khan, A.-H. Mostofa Kamal, Murshid S. M. Hasan, A. Kabir, and D. Yeasmin. "COVID-19-Related Infodemic and Its Impact on Public Health: A Global Social Media Analysis." *American Journal of Tropical Medicine and Hygiene* 103 (2020): 1621–1629.

Joel, S., P. W. Eastwick, and E. J. Finkel. "Is Romantic Desire Predictable? Machine Learning Applied to Initial Romantic Attraction." *Psychological Science* 28 (2017): 1478–1489.

Johnson, B. "Privacy No Longer a Social Norm, Says Facebook Founder." *Guardian*, January 11, 2010. https://www.theguardian.com/technology/2010/jan/11/facebook -privacy.

Jung, J., C. Concannon, R. Shroff, S. Goel, and D. G. Goldstein. "Simple Rules for Complex Decisions." *ArXiv*, February 15, 2017. https://arxiv.org/abs/1702.04690.

Kahneman, D. "Comment on 'Artificial Intelligence and Behavioral Economics.'" In *The Economics of Artificial Intelligence*, edited by A. Agrawal, J. Gans, and A. Goldfarb, 608–610. Chicago: University of Chicago Press, 2019.

Kapoun, J. "Teaching Web Evaluation to Undergrads." *College and Research Libraries News* (August 1998): 522–523.

Katsikopoulos, K., O. Şimşek, M. Buckmann, and G. Gigerenzer. *Classification in the Wild*. Cambridge, MA: MIT Press, 2020.

Katsikopoulos, K., O. Şimşek, M. Buckmann, and G. Gigerenzer. "Transparent Modeling of Influenza Incidence: Big Data or a Single Data Point from Psychological Theory?" *International Journal of Forecasting* (2021).

Kawohl, J. M., and J. Becker. *Verfügen Deutsche Vorstände über die Zukunftsfähigkeit, die die Digitale Transformation Erfordert?* [Can German CEOs Meet the Future Demands of Digital Transformation?]. Heilbronn: Investment Lab, 2017. https:// docs.wixstatic.com/ugd/63eb59_4465a197bb6f4f34b784c3c52e1456fb.pdf.

Kay, J., and M. King. *Radical Uncertainty*. London: Bridge Street Press, 2020.

Kayser-Bril, N. "Facebook Enables Automated Scams, but Fails to Automate the Fight Against Them." *Algorithm Watch*, November 4, 2019. https://algorithmwatch.org /en/story/facebook-enables-automated-scams-but-fails-to-automate-the-fight-against -them/.

Kayser-Bril, N. "Female Historians and Male Nurses Do Not Exist, Google Translate Tells Its European Users." *Algorithm Watch*, September 17, 2020. https://algorithm watch.org/en/story/google-translate-gender-bias/.

Kellermann, A. L., and S. S. Jones. "What It Will Take to Achieve the As-Yet-Unfilled Promises of Health Information Technology." *Health Affairs* 32 (2013): 63–68.

Kim, M. "Are Smartphones Causing More Kid Injuries?" *Philly Voice*, January 8, 2015. https://www.phillyvoice.com/cell-phone-too-deadly-kids/.

Kleinhubbert, G. "Abgelenkt" [Distracted]. *Der Spiegel*, November 25, 2013. https://www.spiegel.de/spiegel/print/d-122579480.html.

Koh, P. W. "WILDS: A Benchmark of In-the-Wild Distribution Shifts." *ArXiv*, March 9, 2021. https://arxiv.org/abs/2012.07421v2.

Kosko, B. "Thinking Machines = Old Algorithms on Faster Computers." In *What to Think about Machines That Think*, edited by J. Brockman, 423–426. New York: Harper, 2015.

Kostka, G. "China's Social Credit System and Public Opinion: Explaining High Levels of Approval." *New Media & Society* 21 (2019): 1565–1593.

Kostka, G., and L. Antoine. "Fostering Model Citizenship: Behavioral Responses to China's Emerging Social Credit Systems." *Policy & Internet* 12 (2019): 256–289.

Kostka, G., and C. Zhang. "Tightening the Grip: Environmental Governance under Xi Jinping." *Environmental Politics* 27 (2018): 769–781.

Kozyreva, A., S. Lewandowsky, and R. Hertwig. "Citizens Versus the Internet: Confronting Digital Challenges with Cognitive Tools." *Psychological Science in the Public Interest* 21 (2020): 103–156.

Kramer, A. D. I., J. E. Guillory, and J. T. Hancock. "Experimental Evidence of Massive-Scale Emotional Contagion through Social Networks." *Proceedings of the National Academy of Sciences* 111 (2014): 8788–8790.

Krauthammer, C. "Be Afraid." *Washington Examiner*, May 26, 1997. https://www.washingtonexaminer.com/weekly-standard/be-afraid-9802.

Kruglanski, A., and G. Gigerenzer. "Intuitive and Deliberate Judgments Are Based on Common Principles." *Psychological Review* 118 (2011): 97–109.

Kumar, N., A. C. Berg, P. N. Belhumeur, and S. K. Nayar. "Attribute and Simile Classifiers for Face Verification." *2009 IEEE 12th International Conference on Computer Vision*, Kyoto (2009): 365–372.

Kurzweil, R. *How to Create a Mind*. New York: Penguin, 2012.

Lake, B. M., T. D. Ulman, J. B. Tenenbaum, and S. J. Gershman. "Ingredients of Intelligence: From Classic Debates to an Engineering Roadmap." *Behavioral and Brain Sciences* 40 (2017): e281.

Lanier, J. "Jaron Lanier Fixes the Internet." *New York Times*, September 23, 2019. https://www.nytimes.com/interactive/2019/09/23/opinion/data-privacy-jaron -lanier.html.

Lanier, J., and E. G. Weyl. "A Blueprint for a Better Digital Society." *Harvard Business Review*, September 26, 2018. https://hbr.org/2018/09/a-blueprint-for-a-better-digital -society.

Lavin, T. "QAnon, Blood Libel, and the Satanic Panic." *New Republic*, September 29, 2020. https://newrepublic.com/article/159529/qanon-blood-libel-satanic-panic.

Lazer, D., R. Kennedy, G. King, and A. Vespignani. "The Parable of Google Flu: Traps in Big Data Analysis." *Science* 343 (March 14, 2014): 1203–1205.

Lea, S., P. Fischer, and K. Evans. *The Psychology of Scams: Provoking and Committing Errors of Judgement.* Report 1070. University of Exeter for the Office of Fair Trading, 2009. https://webarchive.nationalarchives.gov.uk/20140402205717/http://oft.gov.uk /shared_oft/reports/consumer_protection/oft1070.pdf.

Lee, D.-C., A. G. Brellenthin, P. D. Thompson, X. Sui, I.-M. Lee, and C. J. Lavie. "Running as a Key Lifestyle Medicine for Longevity." *Progress in Cardiovascular Diseases* 60, no. 1 (2017): 45–55.

Lee, K.-F. *AI Superpowers: China, Silicon Valley, and the New World Order.* Boston: Houghton Mifflin Harcourt, 2018.

Leetaru, K. "The Data Brokers So Powerful That Even Facebook Bought Their Data— But They Got Me Wildly Wrong." *Forbes*, April 4, 2018. https://www.forbes.com /sites/kalevleetaru/2018/04/05/the-data-brokers-so-powerful-even-facebook-bought -their-data-but-they-got-me-wildly-wrong/#3d727faf3107.

Lewandowsky, S., L. Smillie, D. Garcia, R. Hertwig, J. Weatherall, S. Egidy, R. E. Robertson, et al. *Technology and Democracy: Understanding the Influence of Online Technologies on Political Behaviour and Decision-Making.* EUR 30422 EN. Luxembourg: Publications Office of the European Union, 2020.

Lewis, M. *Moneyball.* New York: W. W. Norton, 2003.

Lewis, R. A., and J. M. Rao. "On the Near Impossibility of Measuring the Returns of Advertising." Paper presented at the AEA Annual Meeting, January 4, 2014. https:// www.aeaweb.org/conference/2014/preliminary.php.

Lewis, R. A., and D. Reiley. "Advertising Effectively Influences Older Users: How Field Experiments Can Improve Measurement and Targeting." *Review of Industrial Organization* 44 (2014): 147–159.

Lichtman, A. J. *Predicting the Next President: The Keys to the White House.* Lanham, MD: Rowman and Littlefield, 2016.

Lin, Z. J., J. Jung, S. Goel, and J. Skeem. "The Limits of Human Prediction of Recidivism." *Science Advances* 6 (2020): eaaz0652.

Lindström, B., M. Bellander, D. T. Schultner, A., P. N. ChangTobler, and D. M. Amodio. "A Computational Reinforcement Learning Account of Social Media Engagement." *Nature Communications* 12 (2021): 1311.

Liptak, A. "Sent to Prison by a Software Program's Secret Algorithms." *New York Times*, May 1, 2017. https://www.nytimes.com/2017/05/01/us/politics/sent-to-prison-by-a-software-programs-secret-algorithms.html.

Locke, J. *An Essay Concerning Human Understanding*. London: Tegg & Son, 1690. Reprint, Oxford: Oxford University Press, 1975.

Lorenz-Spreen, P., S. Lewandowsky, C. R. Sunstein, and R. Hertwig. "How Behavioral Sciences Can Promote Truth, Autonomy and Democratic Discourse." *Nature Human Behavior* 4 (2020): 1102–1109.

Lotan, G. "Israel, Gaza, War, and Data—The Art of Personalizing Propaganda." *Global Voices*, August 4, 2014. https://globalvoices.org/2014/08/04/israel-gaza-war-data-the-art-of-personalizing-propaganda/.

Luckerson, V. "Here's How Facebook's News Feed Actually Works." *Time*, July 9, 2015. https://time.com/collection-post/3950525/facebook-news-feed-algorithm/.

Luetge, C. "The German Ethics Code for Automated and Connected Driving." *Philosophy of Technology* 30 (2017): 547–558.

Luria, A. R. *The Mind of a Mnemonist*. New York: Basic Books, 1968.

Maack, N. "Diese Stadt Ist ein Großer Roboter" [This City Is a Big Robot]. *Frankfurter Allgemeine*, January 22, 2020. https://www.faz.net/aktuell/feuilleton/debatten/toyotas-smart-city-diese-stadt-ist-ein-grosser-roboter-16588460.html.

Makoff, J. "Computer Wins on 'Jeopardy!': Trivial, It's Not." *New York Times*, February 16, 2011. https://www.nytimes.com/2011/02/17/science/17jeopardy-watson.html.

Markman, J. "Big Data and the 2016 Election." *Forbes*, August 8, 2016. https://www.forbes.com/sites/jonmarkman/2016/08/08/big-data-and-the-2016-election/#29fe5d7b1450.

Martosko, D. "I'm Bad at Math, Says Obama—But So Is Congress—During Visit to Brooklyn School." *Daily Mail*, October 25, 2013. https://www.dailymail.co.uk/news/article-2477130/Im-bad-math-says-Obama--Congress--visit-Brooklyn-school.

Matacic, C. "Are Algorithms Good Guides?" *Science* 359 (2018): 263.

McDaniel, B. T. "Parent Distraction with Phones, Reasons for Use, and Impacts on Parenting and Child Outcomes: A Review of the Emerging Research." *Human Behavior and Emerging Technologies* 1 (2019):72–80.

McDonald, A. M., and L. F. Cranor. "The Cost of Reading Privacy Policies." *Journal of Law and Policy for the Information Society* 4 (2008): 543–568.

McGrew, S., J. Breakstone, T. Ortega, M. Smith, and S. Wineberg. "Can Students Evaluate Online Sources? Learning from Assessments of Civic Online Reasoning." *Theory and Research in Social Education* 46 (2018): 165–193.

McGrew, S., M. Smith, J. Breakstone, T. Ortega, and S. Wineburg. "Improving University Students' Web Savvy: An Intervention Study." *British Journal of Educational Psychology* 89, no. 4 (2019): 85–500.

McNeal, G. S. "Facebook Manipulated User News Feeds to Create Emotional Responses." *Forbes*, June 28, 2014. https://www.forbes.com/sites/gregorymcneal/2014/06/28/facebook-manipulated-user-news-feeds-to-create-emotional-contagion/.

Mead, M. *Coming of Age in Samoa*. New York: William Morrow, 1928.

Medvin, M. "Your Vehicle Black Box: A 'Witness' against You in Court." *Forbes*, January 8, 2019. https://www.forbes.com/sites/marinamedvin/2019/01/08/your-vehicle-black-box-a-witness-against-you-in-court-2/#461a443831c5.

Melendez, S. "Mozilla and Creative Commons Want to Reimage the Internet without Ads, and They Have $100M to Do It." *Fast Company*, September 16, 2019. https://www.fastcompany.com/90403645/mozilla-and-creative-commons-want-to-reimagine-the-internet-without-ads-and-they-have-100m-to-do-it.

Mercier, H., and D. Sperber. "Why Do Humans Reason? Arguments for an Argumentative Theory." *Behavioral and Brain Science* 34 (2011): 57–74.

Messerli, F. H. "Chocolate Consumption, Cognitive Function, and Nobel Laureates." *New England Journal of Medicine* 367 (2012): 1562–1564.

Milner, G. "Death by GPS: Are Satnavs Changing Our Brains?" *BBC*, June 25, 2016. https://www.bbc.com/news/technology-35491962.

Mønsted, B., P. Sapiezynski, E. Ferrara, and S. Lehmann. "Evidence of Complex Contagion of Information in Social Media: An Experiment Using Twitter Bots." *PLOS ONE* 12, no. 9 (2017): e0184148.

Montaigne, M. de. *Essays*. Translated by Jonathan Bennett, 2017. https://www.earlymoderntexts.com/assets/pdfs/montaigne1580book2_2.pdf.

Montoya, R. M., R. S. Horton, and J. Kirchner. "Is Actual Similarity Necessary for Attraction? A Meta-Analysis of Actual and Perceived Similarity." *Journal of Social and Personal Relationships* 25 (2008): 889–922.

Morozov, E. *To Save Everything, Click Here: The Follies of Technological Solutionism*. New York: Public Affairs, 2013.

Mullard, A. "Reliability of 'New Drug Target' Claims Called into Question." *Nature Reviews: Drug Discovery* 10, no. 9 (2011): 643–644.

Munro, D. "Should Tech Firms Pay People for Their Data?" Centre for International Governance Innovation, November 28, 2019. https://www.cigionline.org/articles /should-tech-firms-pay-people-their-data.

Nasiopoulos, E., E. F. Risko, T. Foulsham, and A. Kingstone. "Wearable Computing: Will It Make People Prosocial?" *British Journal of Psychology* 106 (2015): 209–216.

National Consortium for the Study of Terrorism and Responses to Terrorism. *Global Terrorism Overview: Terrorism in 2019.* Baltimore: University of Maryland, 2020. https://www.start.umd.edu/pubs/START_GTD_GlobalTerrorismOverview2019_July 2020.pdf.

National Geographic. "Greendex 2014: Consumer Choice and the Environment: A Worldwide Tracking Survey." Globescan, 2014. https://globescan.com/wp-content /uploads/2017/07/Greendex_2014_Full_Report_NationalGeographic_GlobeScan.pdf.

National Transportation Safety Board. *Preliminary Report Highway.* HWY18MH010, 2019. https://www.ntsb.gov/investigations/AccidentReports/Reports/HWY18MH010 -prelim.pdf.

Neumann, N., C. E. Tucker, and T. Whitfield. "Frontiers: How Effective Is Third-Party Consumer Profiling? Evidence from Field Studies." *Marketing Science* 38 (2019): 918–926.

Neuvonen, M., K. Kivinen, and M. Salo, eds. *Elections Approach—Are You Ready? Fact-Checking for Educators and Future Voters.* FactBarEDU, 2018. https://www.faktabaari .fi/assets/FactBar_EDU_Fact-checking_for_educators_and_future_voters_13112018 .pdf.

Newell, A., J. C. Shaw, and H. A. Simon. "Chess-Playing Programs and the Problem of Complexity." *IBM Journal of Research and Development* 2, no. 4 (1958): 320–335.

Newell, A., J. C. Shaw, and H. A. Simon. "Elements of a Theory of Human Problem Solving." *Psychological Review* 65 (1958): 151–166.

Newell, A., and H. A. Simon. *Human Problem Solving.* Englewood Cliffs, NJ: Prentice-Hall, 1972.

Newport, F. "Most U.S. Smartphone Owners Check Phone at Least Hourly." *Gallup*, July 9, 2015. https://news.gallup.com/poll/184046/smartphone-owners-check-phone -least-hourly.aspx.

Newton, C. "A Partisan War over Fact-Checking Is Putting Pressure on Facebook." *The Verge*, September 12, 2018. https://www.theverge.com/2018/9/12/17848478 /thinkprogress-weekly-standard-facebook-fact-check-false.

Nguyen, A., J. Yosinski, and J. Clune. "Deep Neural Networks Are Easily Fooled: High Confidence Predictions for Unrecognizable Images." *2015 IEEE Conference on Computer Vision and Pattern Recognition*, 2015. https://ieeexplore.ieee.org/document /7298640.

Nguyen, N. "If You Have a Smart TV, Take a Closer Look at Your Privacy Settings." *CNBC*, March 9, 2017. https://www.cnbc.com/2017/03/09/if-you-have-a-smart-tv -take-a-closer-look-at-your-privacy-settings.html.

NHS. "Country's Top Mental Health Nurse Warns Video Games Pushing Young People Into 'Under the Radar' Gambling." January 18, 2020. https://www.england .nhs.uk/2020/01/countrys-top-mental-health-nurse-warns-video-games-pushing -young-people-into-under-the-radar-gambling/.

99 Content. "Snapchat Statistics." N.d. https://99firms.com/blog/snapchat-statistics/.

Niven, T., and H-Y. Kao. "Probing Neural Network Comprehension of Natural Language Arguments." *Proceedings of the 57th Annual Meeting of the Association for Computational Linguistics* (2019): 4658–4664. https://www.aclweb.org/anthology /P19–1459.pdf.

Norton LifeLock. *Norton LifeLock Cyber Safety Insights Report: Global Results.* Symantec Corporation, 2019. https://now.symassets.com/content/dam/norton/campaign /NortonReport/2020/2019_NortonLifeLock_Cyber_Safety_Insights_Report_Global _Results.pdf.

Noyes, D. *The Top 20 Valuable Facebook Statistics.* Zephoria Digital Marketing, October 2020. https://zephoria.com/top-15-valuable-facebook-statistics/.

OECD. *Key Issues for Digital Transformation in the G20.* Berlin: OECD, January 12, 2017. https://www.oecd.org/g20/key-issues-for-digital-transformation-in-the-g20.pdf.

Office of Fair Trading, United Kingdom. *Research on Impact of Mass Marketed Scams: A Summary of Research into the Impact of Scams on UK Consumers.* OFT Report 883, 2007. https://webarchive.nationalarchives.gov.uk/20140402214439/http:/www.oft .gov.uk/shared_oft/reports/consumer_protection/oft883.pdf.

Olson, D. R., K. J. Konty, M. Paladini, C. Viboud, and L. Simonsen. "Reassessing Google Flu Trends Data for Detection of Seasonal and Pandemic Influenza: A Comparative Epidemiological Study at Three Geographic Scales." *PLOS Computational Biology* 9, no. 10 (2013): e1003256.

O'Neill, C. *Weapons of Mass Destruction.* London: Allen Lane, 2016.

Ophir, E., C. Nass, and A. D. Wagner. "Cognitive Control in Media Multitaskers." *Procedeedings of the National Academy of Sciences* 106, no. 37 (2009): 15583–15587.

Orben, A. "The Sisyphean Cycle of Technology Panics." *Perspectives on Psychological Science* 15 (2020): 1143–1157.

Orwell, G. *1984*. London: Secker & Warburg, 1949. Reprint, London: Penguin, 2008.

O'Toole, F. *Heroic Failure: Brexit and the Politics of Pain*. London: Head of Zeus, 2018.

Pasquale, F. *The Black Box Society*. Cambridge, MA: Harvard University Press, 2015.

Paul, A. "Is Online Better Than Offline for Meeting Partners? Depends: Are You Looking to Marry or to Date?" *Cyberspychology, Behavior, and Social Networking* 17 (2014): 664–667.

Paulsen, D. "Human Versus Machine: A Comparison of the Accuracy of Geographic Profiling Methods." *Journal of Investigative Psychology and Offender Profiling* 3 (2006): 77–89.

Peteranderl, S. *Under Fire: The Rise and Fall of Predictive Policing*. 2020. https://www .acgusa.org/wp-content/uploads/2020/03/2020_Predpol_Peteranderl_Kellen.pdf.

Pogue, D. "How to End Online Ads Forever." *Scientific American*, January 1, 2016. https://www.scientificamerican.com/article/how-to-end-online-ads-forever/.

Poibeau, T. *Machine Translation*. Cambridge, MA: MIT Press, 2017.

Porter, T. M. *Karl Pearson: The Scientific Life in a Statistical Age*. Princeton, NJ: Princeton University Press, 2004.

Potarca, G. "The Demography of Swiping Right. An Overview of Couples Who Met through Dating Apps in Switzerland." *PLOS ONE* 15, no. 12 (2020): e0243733.

Praschl, P. "Eine Barbie mit WLAN is das Ende der Kindheit" [A Wi-Fi Barbie Means the End of Childhood]. *Welt*, April 22, 2015. https://www.welt.de/kultur /article139863883/.

Quinn, P. C., P. D. Eimas, and M. J. Tarr. "Perceptual Categorization of Cat and Dog Silhouettes by 3- to 4-Month-Old Infants." *Journal of Experimental Child Psychology* 79 (2001): 78–94.

Raji, I. D., and J. Buolamwini. "Actionable Auditing: Investigating the Impact of Public Naming Biased Performance Results of Commercial AI Products." *Conference on Artificial Intelligence, Ethics, and Society*. 2019. https://www.media.mit.edu /publications/actionable-auditing-investigating-the-impact-of-publicly-naming -biased-performance-results-of-commercial-ai-products/.

Ranjan, A. J. Janai, A. Geiger, and M. J. Black. "Attacking Optical Flow." *Proceedings International Conference on Computer Vision (ICCV)* (2019): 2404–2413. https://arxiv .org/abs/1910.10053.

Rebonato, R. *Taking Liberties. A Critical Examination of Libertarian Paternalism*. Basingstoke, UK: Palgrave Macmillan, 2012.

Reed, J., K. Hirsh-Pasek, and R. M. Golinkoff. "Learning on Hold: Cell Phones Sidetrack Parent-Child Interactions." *Developmental Psychology* 53 (2017): 1428–1436.

Reynolds, G. "An Hour of Running May Add 7 Hours to Your Life." *New York Times*, April 12, 2017. https://www.nytimes.com/2017/04/12/well/move/an-hour-of-running -may-add-seven-hours-to-your-life.html.

Rid, T. *Active Measures*. London: Profile Books, 2020.

Rilke, R. M. *Letters to a Young Poet*. Translated by M. D. Herter Norton. New York: Norton, 1993.

Rivest, R. L. "Learning Decision Lists." *Machine Learning* 2, no. 2 (1987): 29–246.

Roberts, S., and H. Pashler. "How Pervasive Is a Good Fit? A Comment on Theory Testing." *Psychological Review* 107 (2000): 58–367.

Rocca, J. "Understanding Generative Adversarial Networks (GANs)." *Towards Data Science*, January 7, 2019. https://towardsdatascience.com/understanding-generative -adversarial-networks-gans-cd6e4651a29.

Rogers, S. "Census in Pictures: From Suffragettes to Arms Protesters." *Guardian*, December 11, 2012. https://www.theguardian.com/uk/datablog/gallery/2012/dec/11 /census-2011-pictures-suffragettes-arms-protesters.

Rompay, T. J. L. van, D. J. Vonk, and M. L. Fransen. "The Eye of the Camera: Effects of Security Cameras on Prosocial Behavior." *Environment and Behavior* 41 (2009): 60–74.

Ross, C., and I. Swetlitz. "IBM Watson Health Hampered by Internal Rivalries and Disorganization, Former Employees Say." *STAT+*, June 14, 2018. https://www .statnews.com/2018/06/14/ibm-watson-health-rivalries-disorganization/.

Rötzer, F. "Jeder Sechste Deutsche Findet ein Social-Scoring System Gut [Every Sixth German Finds the Social Scoring System Good]." *Telepolis*, February 4, 2019. https:// www.heise.de/tp/features/Jeder-sechste-Deutsche-findet-ein-Social-Scoring-System -nach-chinesischem-Vorbild-gut-4297208.html.

Rudin, C. "Stop Explaining Black Box Machine Learning Models of High Stakes Decisions and Use Interpretable Models Instead." *Nature Machine Intelligence* 1 (2019): 206–215.

Rudin, C., and J. Radin. "Why Are We Using Black Box Models in AI When We Don't Need To? A Lesson from an Explainable AI Competition." *Harvard Data Science Review* 1, no. 2 (2019). https://hdsr.mitpress.mit.edu/pub/f9kuryi8/release/6.

Rudin, C., C. Wang, and B. Coker. "The Age of Secrecy and Unfairness in Recidivism Prediction." *Harvard Data Science Review* 2, no. 1 (2020). https://hdsr.mitpress.mit .edu/pub/7z10o269/release/4.

Rudder, C. *Dataclysm*. London: Harper, 2014.

Ruginski, I. T., S. H. Creem-Regeht, J. K. Stefanucci, and E. Cashdan. "GPS Use Negatively Affects Environmental Learning through Spatial Transformation Abilities." *Journal of Environmental Psychology* 64 (2019): 12–20.

Russell, S. *Human Compatible: Artificial Intelligence and the Problem of Control*. New York: Viking, 2019.

Russell, S., and R. Norvig, eds. *Artificial Intelligence: A Modern Approach*. 3rd ed. Upper Saddle River, NJ: Pearson Education, 2010.

Sales, N. J. *American Girls: Social Media and the Secret Lives of Teenagers*. New York: Knopf, 2016.

Salganik, M. J., I. Lundberg, A. T. Kindel, C. E. Ahearn, K. Al-Ghoneim, A. Almaatouq, D. M. Altschul, et al. "Measuring the Predictability of Life Outcomes with a Scientific Mass Collaboration." *Proceedings of the National Academy of Sciences* 117, no. 15 (2020): 8398–8403.

Sarle, W. S. "Neural Networks and Statistical Models." In *Proceedings of the Nineteenth Annual SAS Users Group International Conference*, 1538–1550. Cary, NC: SAS Institute, 1994.

Sattelberg, W. "Longest Snapchat Streak." *TechJunkie*, November 16, 2020. https://social.techjunkie.com/longest-snapchat-streak/.

Schmidt. E. "Google CEO Eric Schmidt on Privacy." December 12, 2009. https://www.youtube.com/watch?v=A6e7wfDHzew

Schüll, N. D. *Addiction by Design*. Princeton, NJ: Princeton University Press, 2012.

Schulte, F., and E. Fry. "Death by 1,000 Clicks: Where Electronic Health Records Went Wrong." *Fortune*, March 18, 2019. https://fortune.com/longform/medical-records/.

Schwertfeger, B. "Künstliche Intelligenz Trifft 'Künstliche Dummheit'" [Artificial Intelligence Meets "Artificial Stupidity"]. Interview with Gert Antes. *wirtschaft + weiterbildung* 9 (2019): 46–49.

Sergeant, D. C., and E. Himonides. "Orchestrated Sex: The Representation of Male and Female Musicians in World-Class Symphony Orchestras." *Frontiers in Psychology* 10 (2019): 1760.

Shahin, S., and P. Zheng. "Big Data and the Illusion of Choice: Comparing the Evolution of India's Aadhaar and China's Social Credit System as Technosocial Discourses." *Social Science Computer Review* 38 (2020): 25–41.

Shladover, S. E. "The Truth about 'Self-Driving' Cars." *Scientific American* 29 (2016): 80–83.

Silver, N. "Election Update: Where Are the Undecided Voters?" *FiveThirtyEight*, October 25, 2018. https://fivethirtyeight.com/features/election-update-where-are-the-undecided-voters/.

Silver, N. "Election Update: Why Our Model Is More Bullish Than Others on Trump." *FiveThirtyEight*, October 24, 2016. https://fivethirtyeight.com/features /election-update-why-our-model-is-more-bullish-than-others-on-trump/.

Simanek, D. E. "Arthur Conan Doyle, Spiritualism, and Fairies." January 2009. https://www.lockhaven.edu/~dsimanek/doyle.htm.

Simon, H. A. *Models of My Life*. New York: Basic Books, 1991.

Simon, H. A. *The Sciences of the Artificial*. Cambridge, MA: MIT Press, 1969.

Simon, H. A, and A. Newell. "Heuristic Problem Solving: The Next Advance in Operations Research." *Operations Research* 6 (1958): 1–10.

Simonite, T. "When It Comes to Gorillas, Google Photos Remains Blind." *Wired*, November 1, 2018. https://www.wired.com/story/when-it-comes-to-gorillas-google -photos-remains-blind/.

Sinha, P., B. J. Balas, Y., Ostrovskyand R. Russell. "Face Recognition by Humans." In *Face Recognition: Advanced Modeling and Methods*, edited by W. Zhao and R. Chellappa, 257–291. London: Academic Press, 2006.

Sinha, P., and T. Poggio. "I Think I Know that Face . . ." *Nature* 384 (1996): 404.

Skinner, B. F. *Beyond Freedom and Dignity*. New York: Bantham, 1972.

Skinner, B. F. *Contingencies of Reinforcement: A Theoretical Analysis*. New York: Appleton-Century-Crofts, 1969.

Smith, B. "We Worked Together on the Internet. Last Week, He Stormed the Capitol." *New York Times*, January 20, 2020. https://www.nytimes.com/2021/01/10 /business/media/capitol-anthime-gionet-buzzfeed-vine.html.

Smith, T., E. Darling, and B. Searles. "2010 Survey on Cell Phone Use while Performing Cardiopulmonary Bypass." *Perfusion* 26 (2011): 375–380.

Snook, B., M. Zito, C. Bennel, and P. J. Taylor. "On the Complexity and Accuracy of Geographic Profiling Strategies." *Journal of Quantitative Criminology* 21 (2005): 1–25.

Snowden, E. *Permanent Record*. New York: Metropolitan Books/Henry Holt, 2019.

Society of Automotive Engineers (SAE). "SAE Issues Update Visual Chart for Its 'Levels of Driving Automation' Standard for Self-Driving Vehicles." December 11, 2018. https://www.sae.org/news/press-room/2018/12/sae-international-releases-updated -visual-chart-for-its-%E2%80%9Clevels-of-driving-automation%E2%80%9D-standard -for-self-driving-vehicles.

Solon, O. "Ex-Facebook President Sean Parker: Site Made to Exploit Human 'Vulnerability.'" *Guardian*, November 9, 2017. https://www.theguardian.com/technology /2017/nov/09/facebook-sean-parker-vulnerability-brain-psychology.

Solsman, J. E. "YouTube's AI Is the Puppet Master of Most of what You Watch." *CNET*, January 10, 2018. https://www.cnet.com/news/youtube-ces-2018-neal-mohan/.

Spinelli, L., and M. Crovella. "How YouTube Leads Privacy-Seeking Users Away from Reliable Information." *Adjunct Publication of the 28th ACM Conference on User Modeling, Adaptation and Personalization* (2020): 244–251. https://dl.acm.org/doi /10.1145/3386392.3399566.

Stadler, R. "Jetzt Wird Abgerechnet" [Now's the Day of Reckoning]. *Süddeutsche Zeitung*, May 9, 2006. https://sz-magazin.sueddeutsche.de/gesellschaft-leben/jetzt -wird-abgerechnet-73071.

Stalder, F. "Paying Users for Their Data." *Mail Archive*, July 24, 2014. https://www .mail-archive.com/nettime-l@mail.kein.org/msg02721.html.

Stern, R. "Self-Driving Uber Crash 'Avoidable,' Driver's Phone Playing Video Before Woman Struck." *Phoenix News*, June 21, 2018. https://www.phoenixnewtimes.com /news/self-driving-uber-crash-avoidable-drivers-phone-playing-video-before-woman -struck-10543284.

Stevenson, M. "Assessing Risk Assessment in Action." *Minnesota Law Review* 103 (2018): 303–384.

Stevenson, P. W. "Trump Is Headed for a Win, Says Professor Who Has Predicted 30 Years of Presidential Outcomes Correctly." *Washington Post*, September 23, 2016. https://www.washingtonpost.com/news/the-fix/wp/2016/09/23/trump-is-headed -for-a-win-says-professor-whos-predicted-30-years-of-presidential-outcomes-correctly /?utm_term=.e3a8b731325c.

Stilgoe, J. "Who Killed Elaine Herzberg?" *OneZero*, December 12, 2019. https:// onezero.medium.com/who-killed-elaine-herzberg-ea01fb14fc5e.

Strickland, E. "How IBM Watson Overpromised and Underdelivered on AI Health Care." *IEEE Spectrum*, April 2, 2019. https://spectrum.ieee.org/biomedical/diagnostics /how-ibm-watson-overpromised-and-underdelivered-on-ai-health-care.

Strimple, Z. "The Matchmaking Industry and Singles Culture in Britain 1970–2000." PhD thesis submitted to University of Sussex, 2017. http://sro.sussex.ac.uk/id /eprint/71609/1/Strimpel%2C%20Zoe.pdf.

Su, J., D. V. Vargas, and K. Sakurai. "One Pixel Attack for Fooling Deep Neural Networks." *IEEE Transactions on Evolutionary Computation* 23 (2019): 828–841.

Suarez-Tangil, G., M. Edwards, C. Peersman, G. Stringhini, A. Rashid, and M. Whitty. "Automatically Dismantling Online Data Fraud." *IEEE Transactions on Information Forensics and Security* 15 (2019): 1128–1137.

Szabo, L. "Artificial Intelligence Is Rushing into Patient Care—and Could Raise Risks." *Scientific American*, December 24, 2019. https://www.scientificamerican.com /article/artificial-intelligence-is-rushing-into-patient-care-and-could-raise-risks/.

Szegedy, C., W. Zaremba, I. Sutskever, J. Bruna, D. Erhan, I. Goodfellow, and R. Fergus. "Intriguing Properties of Neural Networks." In *Proceedings of International Conference on Learning Representations (ICLR)*, 2014. https://arxiv.org/abs/1312.6199.

Taleb, N. N. *The Black Swan*. 2nd ed. New York: Random House, 2010.

Taleb, N. N. *Fooled by Randomness*. New York: Random House, 2001.

Tanz, J. "The Curse of Cow Clicker: How a Cheeky Satire Became a Videogame Hit." *Wired*, December 20, 2011. https://www.wired.com/2011/12/ff-cowclicker/.

Tech Transparency Project. "Google's Revolving Door (US)." April 26, 2016. https://www.techtransparencyproject.org/articles/googles-revolving-door-us.

Tesla. "All Tesla Cars Being Produced Now Have Full Self-Driving Hardware." October 19, 2016. https://www.tesla.com/de_DE/blog/all-tesla-cars-being-produced-now-have-full-self-driving-hardware.

Theile, G. "Parship-Chef im Gespräch: 'Gut Ausgebildete Frauen Haben Es Schwerer'" [Interview with Head of Parship: "It's More Difficult for Highly Educated Women"]. *Frankfurter Allgemeine*, February 14, 2020. https://www.faz.net/aktuell/wirtschaft/parship-chef-ueber-das-flirten-menschen-sollten-sich-mehr-trauen-16632496.html.

Thomas, E. "Why Oakland Police Turned Down Predictive Policing." *Vice*, December 28, 2016. https://www.vice.com/en/article/ezp8zp/minority-retort-why-oakland-police-turned-down-predictive-policing.

Thomas, R. J. "Online Exogamy Reconsidered: Estimating the Internet's Effects on Racial, Educational, Religious, Political and Age Assortative Mating." *Social Forces* 98 (2020): 1257–1286.

Titz, S. "Deep Neural Networks: Researchers Outsmart Algorithms in Order to Understand Them Better." *Horizons*, August 3, 2018. https://www.horizons-mag.ch/2018/03/08/gamers-can-solve-science-problems/.

Todd, P. M., L. Penke, B. Fasolo, and A. P. Lenton. "Different Cognitive Processes Underlie Human Mate Choices and Mate Preferences." *Proceedings of the National Academy of Sciences* 104 (2007): 15011–15016.

Tolentino, J. "What It Takes to Put Your Phone Away." *New Yorker*, April 22, 2019. https://www.newyorker.com/magazine/2019/04/29/what-it-takes-to-put-your-phone-away.

Tomasello, M. *Becoming Human*. Cambridge, MA: Belknap Press of Harvard University Press, 2019.

Tombu, M., and P. Jolicoeur. "Virtually No Evidence for Virtually Perfect Time-Sharing." *Journal of Experimental Psychology: Human Perception & Performance* 30 (2004): 795–810.

Topol, E. *Deep Medicine: How Artificial Intelligence Can Make Healthcare Human Again*. New York: Basic Books, 2019.

Topsell, E. *The History of Four-Footed Beasts and Serpents*. London: E. Cotes, 1658. Reprint, New York: Da Capo Press, 1967.

Tramer, F., et al. "Ensemble Adversarial Training: Attacks and Defenses." *ICLR 6th International Conference on Learning Representations*, 2018. https://arxiv.org/abs/1705 .07204.

Tsvetkova, M., R. García-Gavilanes, L. Floridi, and T. Yasseri. "Even Good Bots Fight: The Case of Wikipedia." *PLoS ONE* 12, no. 2 (2017): e0171774.

Turek, M. *Explainable Artificial Intelligence (XAI)*. Defense Advanced Research Projects Agency (DARPA). https://www.darpa.mil/program/explainable-artificial-intelligence.

Turing, A. "Computing Machinery and Intelligence." *Mind* 54 (1950): 433–460.

Turkle, S. *Reclaiming Conversation*. New York: Penguin Books, 2016.

Twenge, J. M. *iGen*. New York: Atria Books, 2017.

Tzezana, R. "Scenarios for Crime and Terrorist Attacks Using the Internet of Things." *European Journal of Futures Research* 4 (2016): 18.

University of Michigan. "Hacking into Home: 'Smart Home' Security Flaws Found in Popular System." May 2, 2016. https://news.umich.edu/hacking-into-homes-smart -home-security-flaws-found-in-popular-system/.

Vigen, T. *Spurious Correlations*. 2015. https://tylervigen.com/spurious-correlations.

Vodafone Institut für Gesellschaft und Kommunikation. "Big Data: Wann Menschen Bereit Sind, Ihre Daten zu Teilen" [Big Data: When Are People Willing to Share Their Data]. January 2016. https://www.vodafone-institut.de/wp-content/uploads/2016/01 /VodafoneInstitute-Survey-BigData-Highlights-de.pdf.

von Neumann, J. *The Computer and the Brain*. New Haven, CT: Yale University Press, 1958.

Wachter, R. *The Digital Doctor*. New York: McGraw-Hill, 2017.

Wang, A. B. "No, the Government is Not Spying on You Through Your Microwave, Ex-CIA Chef Tells Colbert." *Washington Post*, March 18, 2017. https://www.washing tonpost.com/news/the-switch/wp/2017/03/08/ex-cia-chief-to-stephen-colbert-no -the-government-is-not-spying-on-you-through-your-microwave/

Wang, W., B. Tang, R. Wang, L. Wang, and A. Ye. "Towards a Robust Deep Neural Network in Texts: A Survey." *ArXiv*, 2019. https://arxiv.org/abs/1902.07285.

Ward, A. F., K. Duke, A. Gneezy, and M. W. Bos. "Brain Drain: The Mere Presence of One's Own Smartphone Reduces Available Cognitive Capacity." *Journal of the Association for Computer Research* 2 (2017): 140–154.

Watson, J. M., and D. L. Strayer. "Supertaskers: Profiles in Extraordinary Multitasking Ability." *Psychonomic Bulletin & Review* 17(2010): 479–485.

Watson, L. "Humans Have Shorter Attention Span Than Goldfish, Thanks to Smartphones." *Telegraph*, May 15, 2015. https://www.telegraph.co.uk/science/2016/03/12/humans-have-shorter-attention-span-than-goldfish-thanks-to-smart/.

Webb, A. *The Big Nine*. New York: Public Affairs, 2019.

Wegwarth, O., W. Gaissmaier, and G. Gigerenzer. "Smart Strategies for Doctors and Doctors-in-Training: Heuristics in Medicine." *Medical Education* 43 (2009): 721–728.

Weinshall-Margel, K., and J. Shapard. "Overlooked Factors in the Analysis of Parole Decisions." *PNAS* 108, no. 42 (2011): E833.

Weiser, M. "The Computer for the Twenty-First Century." *Scientific American* (September 1991): 94–104.

Weltecke, D. "Gab es 'Vertrauen' im Mittelalter?" [Did "Trust" Exist in the Middle Ages?]. In *Vertrauen: Historische Annäherungen*, edited by U. Frevert, 67–89. Göttingen: Vandenhoeck & Ruprecht, 2003.

"What to Make of Mark Zuckerberg's Testimony." *Economist*, April 14, 2018. https://www.economist.com/leaders/2018/04/14/what-to-make-of-mark-zuckerbergs-testimony.

Whitty, M. T., and T. Buchanan. "The Online Dating Romance Scam: The Psychological Impact on Victims—Both Financial and Non-Financial." *Criminology and Criminal Justice* (2015): 1–9.

Wineburg, S., and S. McGrew. "Lateral Reading and the Nature of Expertise: Reading Less and Learning More when Evaluating Digital Information." *Teachers College Record* 121, no. 11 (2019): 1–40.

"Women in Computer Science: Getting Involved in STEM." *Computer Science*, September 19, 2021. https://www.computerscience.org/resources/women-in-computer-science/.

Woollett, K., and E. A. Maguire. "Acquiring 'the Knowledge' of London's Layout Drives Structural Brain Changes." *Current Biology* 21, No. 24 (2011): 2109–2114.

Wood, J. "He Made the Mini—and Broke the Mould." *Independent*, July 20, 2005. https://www.independent.co.uk/life-style/motoring/features/he-made-the-mini-and-broke-the-mould-301594.html.

Wu, T. *The Attention Merchants*. London: Atlantic Books, 2016.

Wübben, M., and F. v. Wangenheim. "Instant Customer Base Analysis: Managerial Heuristics Often 'Get It Right.'" *Journal of Marketing*, 72 (2008): 82–93.

Young, R. A., S. K. Burge, K. A. Kumar, J. M. Wilson, and D. F. Ortiz. "A Time-Motion Study of Primary Care Physician's Work in the Electronic Health Record Era." *Family Medicine* 50 (2018): 91–99.

Youyou, W., M. Kosinski, and D. Stillwell. "Computer-Based Personality Judgments Are More Accurate Than Those Made by Humans." *Proceedings of the National Academy of Sciences* 112 (2015): 1036–1040.

Zech, J. R., M. A. Badgeley, M. Liu, A. B. Costa, J. J. Titano, and E. K. Oermann. "Variable Generalization Performance of a Deep Learning Model to Detect Pneumonia in Chest Radiographs: A Cross-Sectional Study." *PLOS Medicine* 15, no. 11 (2018): e1002683.

Zendle D, and P. Cairns. "Video Game Loot Boxes Are Linked to Problem Gambling: Results of a Large-Scale Survey." *PLOS ONE* 13, no. 11 (2018): e0206767.

Zhao, J., T. Wang, M. Yatskar, V. Ordonez, K.-W. Chang. "Men Also Like Shopping: Reducing Gender Bias Amplification Using Corpus-Level Constraints." *Proceedings of the 2017 Conference on Empirical Methods in Language Processing* (2017): 2979–2989. https://www.aclweb.org/anthology/D17–1323.pdf.

Zimmerman, F. J., D. A. Christakis, and A. N. Meltzoff. "Associations Between Media Viewing and Language Development in Children Under Age 2 Years." *Journal of Pediatrics* 151 (2007): 364–368.

Zuboff, S. *The Age of Surveillance Capitalism*. London: Profile Books, 2019.

Zuboff, S. "Surveillance Capitalism and the Challenge of Collective Action." *New Labor Forum* 28 (2019): 10–29.

Zuboff, S. "You Are Now Remotely Controlled." *New York Times*, January 24, 2020. https://www.nytimes.com/2020/01/24/opinion/sunday/surveillance-capitalism .html.

Zuckerberg, M. "We Need to Rely on and Build More AI Tools to Help Flag Certain Content." *CNBC*, April 12, 2018. https://www.youtube.com/watch?v=riFuznlZvyY.

Zuckerman, E. "The Internet's Original Sin." *Atlantic*, August 14, 2014. https://www .theatlantic.com/technology/archive/2014/08/advertising-is-the-internets-original -sin/376041/.

Zweig, K. A. "Watching the Watchers: Epstein and Robertson's 'Search Engine Manipulation Effect.'" *Algorithm Watch*, April 7, 2017. https://algorithmwatch.org /en/watching-the-watchers-epstein-and-robertsons-search-engine-manipulation -effect/.

Index

Page numbers in italics indicate figures.

NASA, 23, *45*
National Public Radio (NPR), 221
Navigating, and GPS, 197–198
Netherlands, privacy paradox in, 146,
  147
Neural networks, 48, 57
  artificial neural networks, 55
  convolutional neural networks (CNN),
    71–72
  deep artificial neural networks, 54–56
  and discrimination by race and
    gender, 124–125
  errors made by, 57–62
  and hard-negative mining, 237n17
  input layer, 55
  and language translation, 77–79,
    80–81
  object recognition, 87–89
  output layer, 55
  and random noise, 85–87
  and the Russian tank fallacy, 70–72
  scenario recognition, 89–90
  and transparency, 136
  and warrant, 79
Newell, Allan, 21, 46–47
News Feed (Facebook), 179
Newspeak, 148, 167, 170
New Yorker, The (magazine), xvi, 164
New York Times, 113, 149, 218–219,
  221
New Zealand, privacy paradox in, 146,
  147
1984 (Orwell), 202–203
Nix, Alexander, 163–164
Nobel Memorial Prize in Economic
  Sciences, 21
Nobel Prize and chocolate consumption
  correlation, 99–101
Noise, random: and object recognition,
  85–87
Norton LifeLock, 146
"Nose Dive" (episode of *Black Mirror*),
  139–140, 141

Notifications (Facebook), 180
Nudging, 160–162

Oakland A's, 210–211
Oakland Police Department, 121
Obama, Barack, 131, 134–135, 152
Objectivity, checking online sources
  for, 222
Object recognition, 81, 87–88
Observer, 149–150
Ohio, and self-driving cars, 50
OkCupid, 3, 9–10, 11, 16. *See also*
  Online dating
Olympics, Beijing, 141
One-good-reason family of algorithms,
  131
Online dating and love algorithms, 4–9
  and arranged marriages, 17–19
  and changes in customers' behavior,
    11–12
  profiles in, 9–11
  scams in, 13–17
Online sources, evaluating, 222–223
Opacity, faith in, 128–130
Operant conditioning, 173–174, 249n1.
  See also Skinner boxes
Operation LASER, 120
Optimizing, 12
Oracle Data Cloud, 105, 143
Orchestras, philharmonic, 123
  blind auditions for, 122
Orwell, George, 148, 202–203

Page, Larry, 80, 151
Parents, distracted, 191–195
Parker, Sean, 178, 192
Parole, secret algorithms and, 114
Parship, 3–4. *See also* Online dating
Paternalism, technological, xx–xxi
Pay-as-you-drive algorithm, 131
Payback, 143
Pay-for-service, business model for
  social media, 155–156